KU-120-867

THE BOOK OF WARPLANES

THE BOOK OF WARPLANES

By Kenneth Munson
Associate RAeS

Illustrated by John W Wood
and Associates

CRESCENT BOOKS
NEW YORK

Published 1985 by Crescent Books,
distributed by Crown Publishers, Inc.,
One Park Avenue. New York.
N.Y. 10016

ISBN 0 517 469502

h g f e d c b a

Typeset by Poole Typesetting Ltd

Printed in Hong Kong

CONTENTS

PREFACE AND ACKNOWLEDGEMENTS

This book brings together the best military material from the fourteen volume *Pocket Encyclopaedia of World Aircraft in Colour,* published by Blandford Press between 1966 and 1977. This series enjoyed immense popularity and success – well over 1,000,000 copies were sold in nine languages. The selection in this book covers the principal military aircraft from the primitive reconnaissance, bomber and fighter aircraft of World War I to the sophisticated missile-armed, radar-guided, nuclear weapon-carrying aircraft of today.

Any selection for a volume of this size must necessarily be an arbitrary one. Ours was made upon the most important or representative aircraft, on an international basis, in each period of military aviation's history. Military helicopters, though no less deserving of consideration, have been omitted only for reasons of space. It is hoped that this book will provide a 'ready-reckoner' for the aviation enthusiast, and that the many marking and camouflage schemes, many infrequently illustrated in colour, will provide inspiration for the aircraft modeller. Several of the aircraft are presented in periods of service or in the markings of air forces which may be unfamiliar, but it is hoped that this will lend depth to the perspective.

The **colour plates** are arranged in approximate chronological order of the aircraft's entry into service. The 'split' plan view depicts the markings appearing above and below either the port side or the starboard side of the aircraft, according to the aspect shown in the side elevation. It should not be assumed, however, that the unseen portion of the plan view is necessarily a 'mirror image' of the half portrayed: for example, on post-1941 US aircraft, national insignia normally appear only on the port upper and starboard lower wing surfaces. Neither should it be assumed that all colour plates necessarily show either a standard colour scheme or a pristine ex-works finish; indeed, the intention has been to illustrate a wide variety of finishes, from ex-works to much-weathered aircraft. The **specifications** relate to the specific machine illustrated, and may not necessarily apply in all details to the type or sub-type in general.

Preparation of the colour plates owes a tremendous amount to Ian Huntley, whose comprehensive knowledge of aircraft colours and markings, based upon extensive researches, and whose general advice, formed the foundation upon which the plates were based. The plates were executed, under art editor and director John W Wood, by Michael Barber, Norman Dinnage, Frank Friend, Brian Hiley, Bill Hobson, Alan ('Doc') Holliday, James Jessop, Tony Mitchell, Jack Pelling and Allen Randall.

For other reference material I am indebted to items published at various times by *The Aeromodeller; Aeroplane Monthly; Air Classics; Air Pictorial; Air Progress; AiReview; Aviation Magazine International; Flying Review International; Interconair;* the magazines of IPMS Australasia, Canada, France, UK and USA; the *Journals* of the American Aviation Historical Society, and Cross & Cockade (Great Britain); Harleyford Publications Ltd; Profile Publications Ltd; and *Markings and Camouflage Systems of Luftwaffe Aircraft in World War II* (4 vols) by Karl Ries Jr. Individual material or other assistance, also greatly appreciated, was given by Chaz Bowyer, J M Bruce, Mário Roberto vaz Carneiro, the late Peter L Gray, Lt Col Nils Kindberg, W B Klepacki, Marcelo W Miranda, Douglas Pardee, Stephen Peltz, Eiichiro Sekigawa, Gordon Swanborough, John W R Taylor and John Wickenden. To them all, my grateful thanks.

K M

Seaford, East Sussex
1981

Fokker E.III

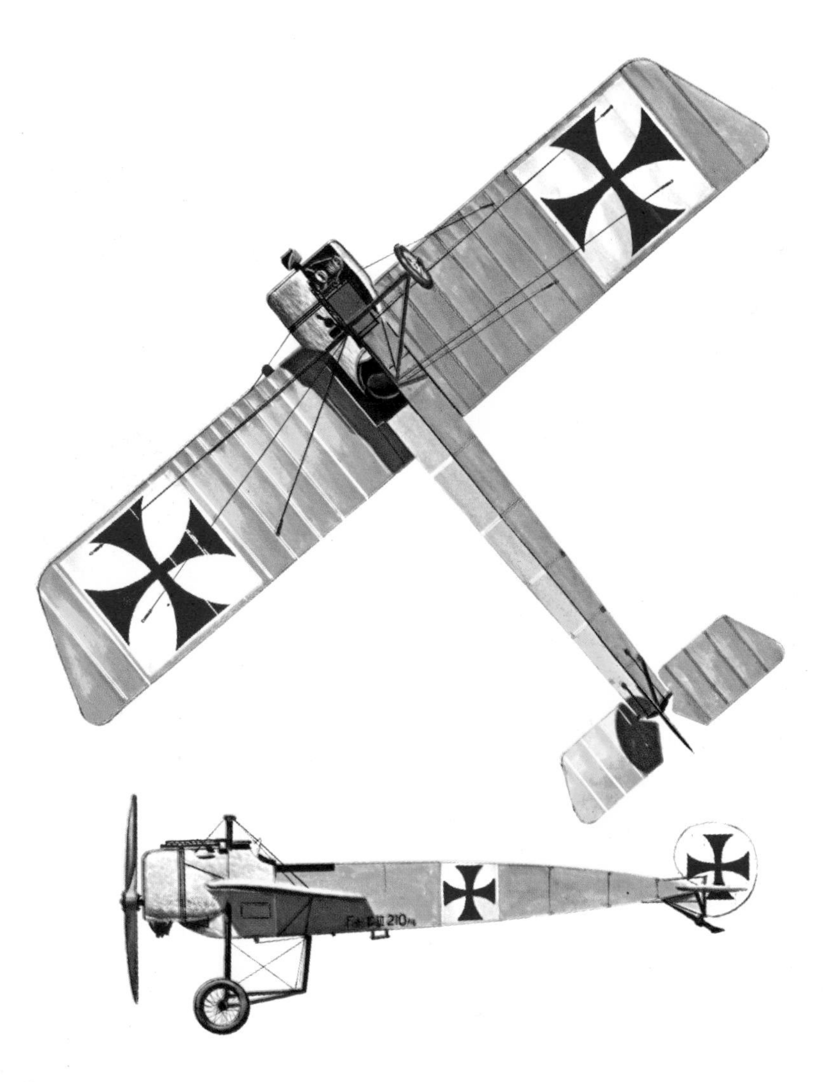

Fokker E.III (210/15) which force-landed behind British lines, 8 April 1916; now in the possession of the Science Museum, London

Span: 31ft 2¾in (9.52m)
Length: 23ft 11⅓in (7.30m)
Weight: 1,400lb (635kg)
Engine: 100hp Oberursel U.I 7-cyl rotary
Max speed: 83mph (134kmh) at 6,500ft (1,980m)
Operational ceiling: 11,500ft (3,500m)
Endurance: 2hr 45mins
Armament: 1 or 2×7.9mm Spandau machine-guns

The Fokker M5 which gave rise to the E types first flew in 1913 and its design was influenced by the Morane-Saulnier Type H monoplane. From it were developed the long-span M5L and short-span M5K, both powered by the 80hp Oberursel rotary and used in modest numbers early in the war; a few M5Ls served with the Austro-Hungarians. Following the capture of Roland Garros' Morane-Saulnier Type L on 19 April 1915, which had deflector plates on the propeller to allow the gun to fire through the arc, Fokker engineers developed an interrupter gear that was tested on three M5Ks using a Spandau gun and this version went into urgent production as the E.I. E.Is were delivered from June 1915, and production probably totalled 68. The E.I, essentially a 'rush job', was quickly followed by the E.II with a 100hp Oberursel and a reduction in wing size which made it more difficult to fly without any appreciable increase in performance. Deliveries, of about 50 E.IIs, began in July 1915. The most numerous version was the E.III, with wings of greater span than the E.II. The E.III was normally fitted with a 7.92mm Spandau gun; a few were flown with twin guns, but the extra weight imposed an unacceptable penalty on performance. Some 260 E.IIIs are thought to have been completed, being used by the German Army and the Austro-Hungarian Air Service, which also received a few E.Is. An attempt to produce a viable 2-gun fighter resulted in the E.IV, whose prototype appeared in November 1915. This had a 160hp two-row Oberursel, and wing span extended to 32ft 9⅔in (10m). However, the E.IV, although faster than the E.III, was less manoeuvrable, and only about 45 were built. Max Immelmann briefly flew a 3-gun E.IV, but reverted to a standard 2-gun machine. On E types serving with the Austro-Hungarian forces the domestic 8mm Schwarzlose gun was the usual weapon. E types were in service on the Western Front from mid-1915 until late summer 1916. The first victory fell to Immelmann on 1 August. Their peak period of effectiveness began about October. From then until January they were virtually unopposed in European skies, leading to the legendary 'Fokker scourge', out of all proportion to their numbers. Their chief victim was the BE2c, which earned a reputation as 'Fokker fodder'. From January 1916 the Fokkers began to encounter worthier opposition – the DH2, Nieuport Nie 11 and FE2b, which virtually dispelled the menace by spring. E types served until the end of 1916 on the Eastern Front and in Mesopotamia, Palestine and Turkey, thereafter being relegated to training.

Airco DH2

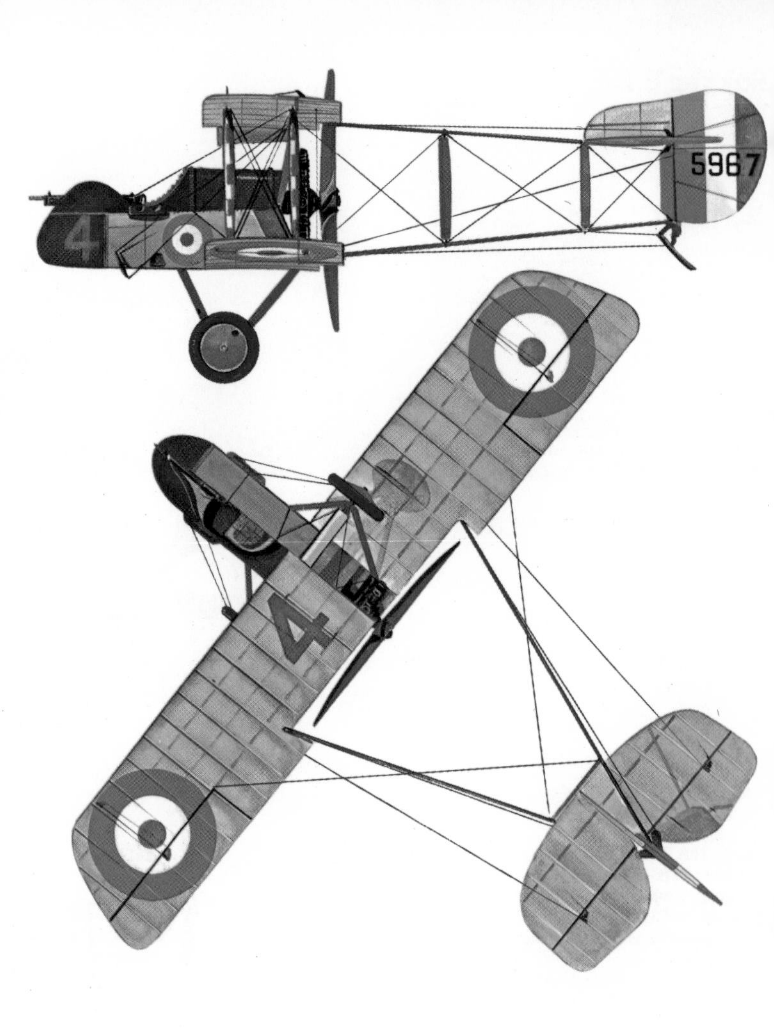

Airco-built DH2 (5967) of A Flight, No 24 Sqdn, RFC, Bertangles, spring 1916

Span: 28ft 3in (8.61m)
Length: 25ft 2½in (7.68m)
Weight: 1,441lb (654kg)
Engine: 100hp Gnome Monosoupape 9-cyl rotary
Max speed: 93mph (150kmh) at sea level
Operational ceiling: 14,500ft (4,420m)
Endurance: 2hr 45min
Armament: 1×0.303in Lewis machine-gun

The DH2 was essentially a scaled-down version of the 2-seat DH1/1A. It was a single-seat pusher whose prototype, 4732, appeared in spring 1915 powered by a 100hp Gnome Monosoupape. A Lewis gun was installed on a pivot mounting to the port side of the cockpit. In July 1915 it was sent to France for operational trials with No 5 Sqdn, RFC, but was brought down in enemy territory on 9 August. However, the DH2 was ordered for quantity production and the first deliveries were made in late 1915 to No 24 Sqdn. This unit went to France in February 1916 with 12 DH2s, the first-ever squadron to be fully equipped with single-seat fighters. The Lewis was now situated centrally in the prow, on a free mounting that enabled it to be traversed or elevated. Pilots often secured it in a forward firing position so that, by aiming the aircraft, they aimed the gun. They found the DH2 tricky to handle at first but when they had mastered it discovered that it was rugged and highly manoeuvrable. The Monosoupape was retained, despite its propensity for shedding cylinders. The DH2 possibly did more than any other Allied aircraft to overcome the Fokker threat. No 24 Sqdn scored its first victory on 2 April 1916 and claimed its first Fokker on 25 April, after which crews attacked the Fokkers with gusto. Later that spring Nos 29 and 32 Sqdns were fully equipped with DH2s, and other squadrons to use them included Nos 5, 11 and 18. About two-thirds of the 400 DH2s built were sent to France. Major L W B Rees, No 32 Sqdn's CO, was awarded the VC for his single-handed attack in a DH2 on 10 German 2-seaters on 1 July 1916. By autumn the DH2 was being outclassed by new German biplane fighters, and on 23 November 1916 Major L G Hawker, VC, DSO, the CO of No 24 Sqdn, was shot down by von Richthofen flying an Albatros D.III. Withdrawal of the DH2 from France started in March 1917, and was completed in June. DH2s serving with the 5th Wing, No 11 Sqdn and X Flight in Palestine, and with A Flight of No 47 Sqdn and a joint RFC/RNAS squadron in Macedonia, had a longer operational life. The Admiralty did not adopt the type. After withdrawal from front-line duties, many were allocated to training units in the UK, where they served until 1918.

Nieuport 17

Nieuport 17C.1, believed to be an aircraft of No 1 Sqdn, RFC, France, *circa* April 1917

Span: 26ft 11⅝in (8.22m)
Length: 18ft 10in (5.74m)
Weight: 1,246lb (565kg)
Engine: 110hp he Rhône 9J 9-cyl rotary
Max speed: 110mph (177kmh) at 6,560 ft (2,000m)
Operational ceiling: 17,390ft (5,300m)
Endurance: 2hr
Armament: 1×0.303in Lewis and 1× 0.303in Vickers machine-guns

The Nieuport 17 retained the 110hp Le Rhône used in the Nieuport 16, differing chiefly from earlier Nieuports in having wings of increased span and area, and was often called the Nieuport '15' or '15-metre', the approximate area of its wings. The Nie 11's single-spar lower wings had been liable to twist when the aircraft was dived or manoeuvred tightly, and a stronger spar was therefore fitted to the Nie 17. In service the Nie 17 became one of the most successful and popular fighting aircraft of the war, both for flying qualities and fighting ability. It was flown by premier pilots including Bishop, Boyau, Guynemer and Nungesser – its own recommendation. It had a fine view from its cockpit, was a first-class dog-fighter and in the words of Cdr C R Samson, RNAS, climbed 'like a witch'. It also had a reputation for 'balloon-busting' with Le Prieur rockets. Early Nie 17s had a Lewis gun on a Foster mounting over the top centre-section, but a single synchronised Vickers replaced this on later machines. Twin-gun installations imposed an unacceptable performance penalty. First French unit to receive the Nie 17 was *Escadrille N57* on 2 May 1916; other French *escadrilles* included *N3, N38, N55, N65* and *N103.* British Nie 17s were in service within weeks, eventually serving with Nos 1, 2, 3, 6, 8, 9, 10 and 11 Sqdns RNAS and Nos 1, 29, 32, 40 and 60 Sqdns RFC on the Western Front. The type was also used by the RFC in Macedonia. The number in British service is indeterminate: only 89 known serial numbers (all for RFC machines) apply specifically to Nieuport 17s, but this was a fraction of the overall total. One authority quotes 527 Nieuport 11/17s in RFC/RNAS service, mostly 17s. Macchi in Italy built 150 Nieuport 17s; it was built in and served in Russia; 20 were supplied to the Dutch Army Air Service in 1917; others to the Belgian *Aviation Militaire* and two to Finland. The AEF purchased 76, as pursuit trainers. In August 1917 there were still 317 Nieuport 17s in front-line French service. Late-production aircraft with 130hp Clerget 9B engines were designated Nieuport 17*bis*. The Nieuport 21 was a 2-seater trainer conversion with an 80hp Le Rhône: 198 were sold to the USA in 1917-18. The Nieuport 23 had improved streamlining and tail surfaces similar to those that appeared later on the Nieuport 28. Some Nieuport 23s were sold after the war to the Swiss *Fliegertruppe* and several single- and two-seaters became sporting or privately owned aircraft postwar. The success of the Nieuport fighters, and the Nie 17 in particular, was maintained even against the theoretically superior Albatros DI and early Halberstadt fighters, and it is no small tribute to the French machines that later German fighters were designed on instructions that they should incorporate many of the features that had made the Nieuports so outstanding.

Sopwith Pup

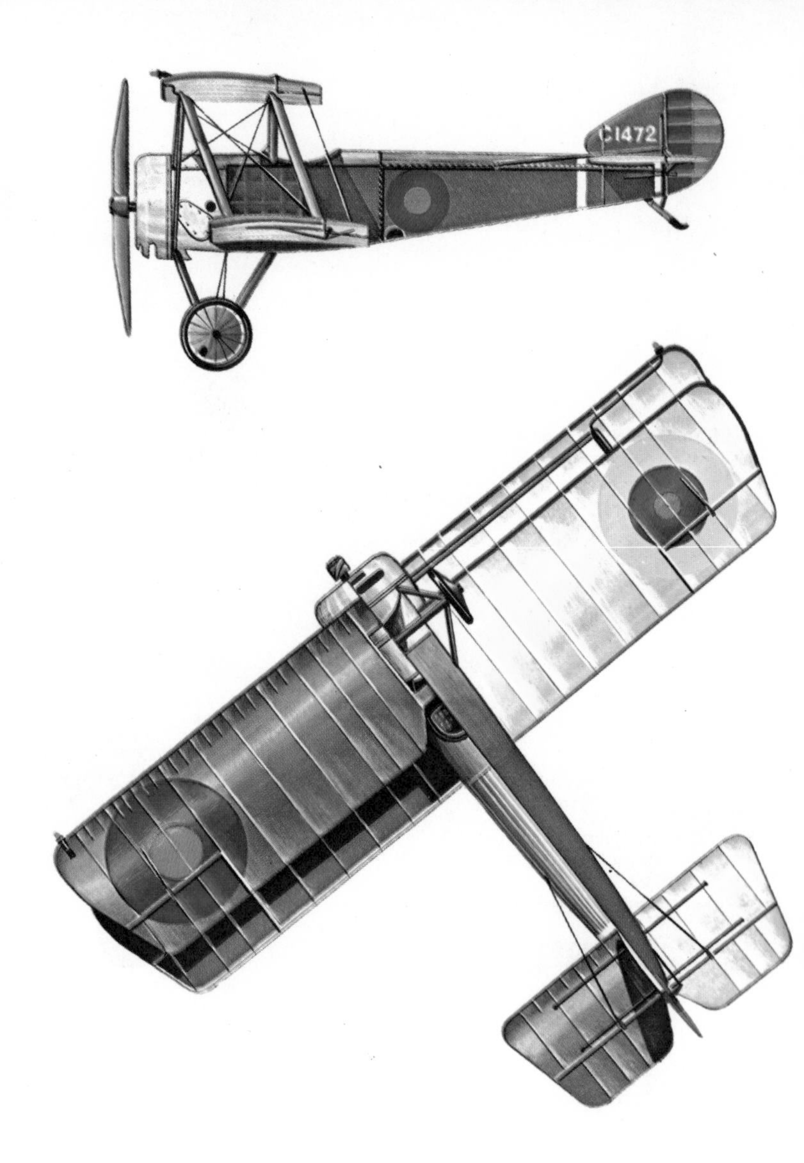

Whitehead-built Pup (unarmed) used for night fighter camouflage trials at Orfordness, March 1918

Span: 26ft 6in (8.08m)
Length: 19ft 3¾in (5.89m)
Weight: 1,225lb (556kg)
Engine: 80hp Le Rhône 9C 9-cyl rotary
Max speed: 111.5mph (179kmh)
Operational ceiling: 17,000ft (5,180m)
Endurance: 3hr
Armament: 1×0.303in Vickers machine-gun. (NB: aircraft illustrated is unarmed)

Developed by Herbert Smith in 1915, the Pup looked like a scaled-down offspring of the earlier 1½-Strutter: hence its nickname, which persisted in spite of Admiralty orders to the contrary. The Pup was classically simple in appearance and construction, and had flying qualities described as 'perfect' and 'impeccable'; but it was rugged, and for 80hp the Pup offered remarkable performance and manoeuvrability. Although the bulk of later production was to RFC orders, the Pup's origins and later associations lay mainly with the RN. The first of six prototypes was flown in February 1916, and the type underwent Admiralty service trials in May. Initial orders were placed with Sopwith and Beardmore. The Le Rhone 9C engine became standard. Officially Admiralty Type 9901, the RNAS Pup was armed with one synchronised 0.303in Vickers gun. Eight Le Prieur rockets were a little-used alternative armament. Deliveries began in August/September 1916, the first recipient being No 1 Wing RNAS. Beardmore and Sopwith built 170 Pups for the Navy, while Whitehead Aviation and the Standard Motor Co completed 1,670 for the RFC. Pups operated on the Western Front in late 1916/early 1917; in October 1916 Naval 8 was formed to assist the RFC. In the two final months of 1917 the Pups of No 8 accounted for 20 enemy aircraft. From the beginning of 1918, although production had still not reached its peak, the Pup began to be withdrawn from front-line units; large numbers were transferred to training establishments, where they were extremely popular. During 1917, several Pups were modified for shipboard anti-Zeppelin duties, with a tripod-mounted Lewis in front of the cockpit, firing upward. The first Pup to land on an aircraft carrier (HMS *Furious*), on 2 August 1917, was flown by Sqdn Cdr E H Dunning, but he was killed later attempting his third landing-on. From July 1917 Pups also served with three RFC and five RNAS Home Defence squadrons. However, they had neither the range nor altitude to deal satisfactorily with German bombers. Altogether, Pups were carried by 5 aircraft carriers and 7 RN cruisers during 1917-18. A Pup development with folding, unstaggered wings and slightly longer fuselage was evolved by Beardmore as the WB.III: 100 were built but their handling qualities were inferior to the standard Pup. After the war the Pup disappeared quickly from British aviation. Eight were placed on the civil register. Eleven were supplied to the RAAF in 1919, and small numbers went to the USA, Greece, the Netherlands and Russia.

Spad VII

British Blériot and Spad Co-built Spad VII of No 23 Sqdn, RFC, La Lovie, autumn 1917

Span: 25ft 7¾in (7.82m)
Length: 20ft 2⅛in (6.15m)
Weight: 1,554lb (705kg)
Engine: 150hp Hispano-Suiza 8Aa 8-cyl V-type
Max speed: 119mph (191.5kmh) at 6,560ft (2,000m)
Operational ceiling: 17,500ft (5,335m)
Endurance: 2hr 15min
Armament: 1×0.303in Vickers machine-gun

In 1915, foreseeing that the rotary engine was reaching the limit of its development, Marc Birkigt, Swiss-born chief designer of Hispano-Suiza, developed a new water-cooled V8 stationary engine promising an initial 150hp. Around such an engine, Louis Bêchereau produced a tractor biplane known as the Spad S V in 1915. From this, Bêchereau developed the Spad S VII, whose prototype flew at Villacoublay in April 1916 powered by a 140hp Hispano-Suiza; it was armed with a forward-firing Vickers gun, offset slightly to starboard, with synchronising gear also designed by Birkigt. The French authorities immediately ordered 268 S VIIs. Delivery began on 2 September 1916, and ultimately 5,600 S VIIs were built in France by eight manufacturers. Early aircraft had the 150hp Hispano-Suiza 8Aa, later models being given increased wing span and rudder area as the 180hp and 200hp models became available. The British Blériot and Spad Co built 100, and Mann, Egerton built 120, for the RFC and RNAS respectively but the RNAS exchanged its S VIIs for Sopwith triplanes on order for the RFC. Although less manoeuvrable than the Nieuports, the S VII was a strong, stable gun platform, with a first-rate turn of speed and an excellent climb to 12,000ft (3,660m). It filled a dire need when the British air forces in particular were equipped with vulnerable pusher types. On the Western Front it served with numerous *escadrilles de chasse*, including the famous *SPA3, Les Cigognes*. From October 1916 it equipped Nos 19 and 23 Sqdns RFC, and 15 were supplied to *Escadrilles 5* and *10* of the Belgian *Aviation Militaire*. Another 19, some fitted in the field with a wing-mounted Lewis gun in addition to the Vickers, were supplied to three RFC squadrons in Mesopotamia, and others went to training units in the UK. Italy was supplied with 214 S VIIs which equipped five *squadriglie* including the celebrated 91ª commanded by Francesco Baracca; a number were delivered to Russia, where they sometimes carried Le Prieur rockets in addition to gun armament; and in December 189 S VIIs were bought by the USA, which allocated a proportion of them to seven squadrons of the AEF in Europe and sent the remainder home to serve as trainers. French squadrons began to re-equip with S XIIIs during mid-1917. The S XII, inspired by Guynemer, was virtually an S VII with a single-shot 37mm gun mounted between the engine-blocks. After the war about 100 S VIIs, many of them rebuilds, were supplied to the *Ecole Blériot* at Buc, and many others were sold to air forces all over the world.

Royal Aircraft Factory RE8

Daimler-built RE8 of A Flight, No 59 Sqdn, RAF, Vert Galand, May 1918

Span: 42ft 7in (12.98m)
Length: 27ft 10½in (8.50m)
Weight: 2,678lb (1,215kg)
Engine: 150hp RAF 4a 12-cyl V-type
Max speed: 102mph (164.2kmh) at 6,500ft (1,980m)
Operational ceiling: 13,500ft (4,115m)
Endurance: 4hr 15min
Armament: 1×0.303in Vickers, and 1× 0.303 Lewis machine-guns
Max bomb load: 224lb (102kg)

The RE8, developed late in 1915, was to provide the RFC with a better defended replacement for the BE2. It certainly carried a more effective armament, but the stubborn adherence by the Factory to the inherent stability concept meant that the RE8, like its predecessors, could easily be outmanoeuvred by German fighters. Two prototypes (7996 and 7997) were completed, the first flying on 17 June 1916, powered by RAF 4a engines, the standard installation in production RE8s; they had BE2e-type wings with marked stagger and dihedral. Production began in August 1916, the first few machines having, like the prototypes, a drum-fed Lewis gun mounted low down in the cockpit, firing between propeller blades fitted with bullet deflector plates. A ring-mounted Lewis was provided in the rear cockpit. Standard front armament of the RE8 soon became a synchronised Vickers gun, mounted under the port-side engine panels. Deliveries began in November 1916, the first aircraft arriving in France that month with No 56 Sqdn RFC. The inexplicable reduction in fin area on production RE8s resulted in several being lost in spinning and other accidents or from fires after crash-landings. As a result, the upper and lower fin areas were enlarged. The RE8 was the most widely used British 2-seater on the Western Front, 4,099 being completed by seven British manufacturers. These included 22 supplied to *Escadrille 6* of the Belgian *Aviation Militaire.* RE8s equipped 16 RFC/RAF squadrons in France, and the type was in service throughout 1917-18. Their duties included observation, reconnaissance, ground-support patrols, night bombing with two 112lb bombs or equivalent smaller bombs, and ground attack with four 65lb bombs. Despite being outclassed by enemy fighters, they served well for much longer than should have been necessary, and 15 RAF squadrons had RE8s at the time of the Armistice. They were used by two British squadrons in Italy, two in Mesopotamia and four in Palestine; three Home Defence units also had some RE8s. A few remained in RAF service for a short time after the war. Variants included the RE8a, with a 200hp Hispano-Suiza and Vickers gun on top of the front fuselage; it is thought only one was built. The RE9 and RT1 were both potential RE8 replacements, utilising many of its components. Several RE8s were converted to RE9s with equal-span wings and modified control surfaces; Siddeley-Deasy built six RT1s with a variety of powerplants, unequal-chord wings and a Lewis gun over the top plane.

Airco DH4

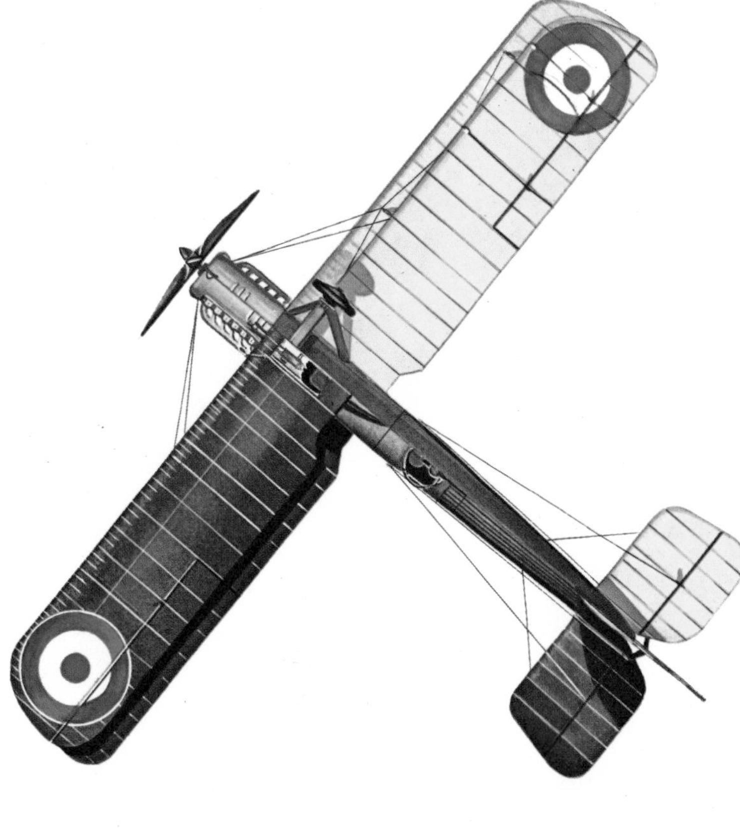

Westland-built DH4 of No 5 Wing, RNAS, Dunkirk, *circa* summer 1917

Span: 42ft 4⅝in (12.92m)
Length: 30ft 8in (9.35m)
Weight: 3,313lb (1,503kg)
Engine: 250hp Rolls-Royce III (Eagle III) 12-cyl V-type
Max speed: 117mph (188.3kmh)
Operational ceiling: 16,000ft (4,875m)
Endurance: 3hr 30min
Armament: 2×0.303in Vickers, and 1× 0.303in machine-guns
Max bomb load: 460lb (209kg) (2× 230lb or 4×112lb bombs)

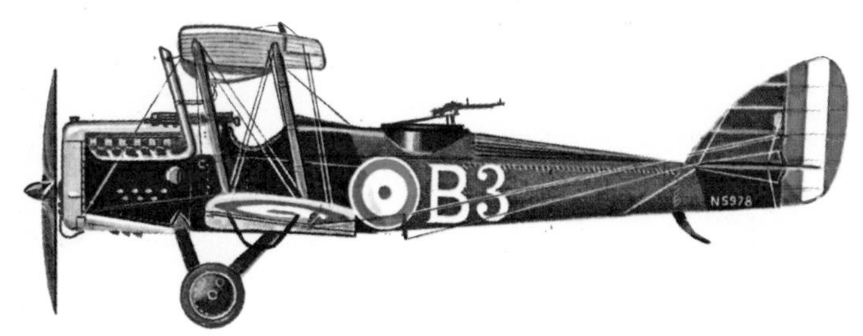

The DH4 was the first British aeroplane designed specifically for high-speed day bombing, although it was also employed on other duties. It was designed by Geoffrey de Havilland around the 160hp BHP, which powered the first prototype when it flew in August 1916. Fifty DH4s were ordered from Westland with 250hp Rolls-Royce Eagle III or IV engines for the RNAS; they had twin Vickers front guns and an elevated Scarff ring to improve the observer's field of fire. Comfortable, light on the controls and easy to fly, the DH4's main drawback was that the fuel between the cockpits was vulnerable to gunfire and inhibited communications between the pilot and observer. DH4 production outstripped Rolls-Royce engine supply; alternatives included 200hp Puma, Adriatic or RAF 3a, 260hp Fiat A-12 and 375hp Eagle VIII engines. The Eagle DH4 had a superlative performance, including a maximum speed of 143mph. Later production aircraft had larger-diameter propellers and longer undercarriage legs. First RFC deliveries were to No 55 Sqdn, which made its first raid in France in April 1917. The first RNAS DH4 unit, No 2 Sqdn, became operational at the same time. DH4s served with six RFC/RAF and five RNAS/RAF squadrons. The RNAS made more varied use of its DH4s' ability to out-fly and climb above fighters, operating without escort. They were used for bombing, artillery spotting, anti-submarine and photo-recce. RFC or RNAS DH4s operated in Russia, Macedonia, Mesopotamia, the Aegean and the Adriatic, and at home for coastal patrol and training. Withdrawn from RAF service soon after the Armistice, several were supplied to Belgium, Canada, Chile, Greece, Iran, New Zealand, South Africa, and Spain, serving with some of these countries until the early 1930s. Seven British manufacturers built 1,499 DH4s; SABCA in Belgium built 15 in 1926. Largest DH4 production was in the USA, and it was the only American-built warplane in combat service in World War 1. A pattern aircraft was sent to America in July 1917, and fitted with a 400hp Liberty 12, flying on 29 October. Designated 'Liberty Planes', 4,846 DH-4As were built. Contracts for 7,502 were cancelled after the Armistice; only 30 per cent reached France, serving with 17 US squadrons from August 1918. The Americans were never happy with the semi-obsolete DH-4A. The improved DH-4B appeared too late to serve in France, but a considerable DH-4A conversion programme kept the DH-4B, DH-4M and other variants in US service until 1932; 283 DH-4As were transferred to the USN and USMC during and after World War 1.

Royal Aircraft Factory SE5A

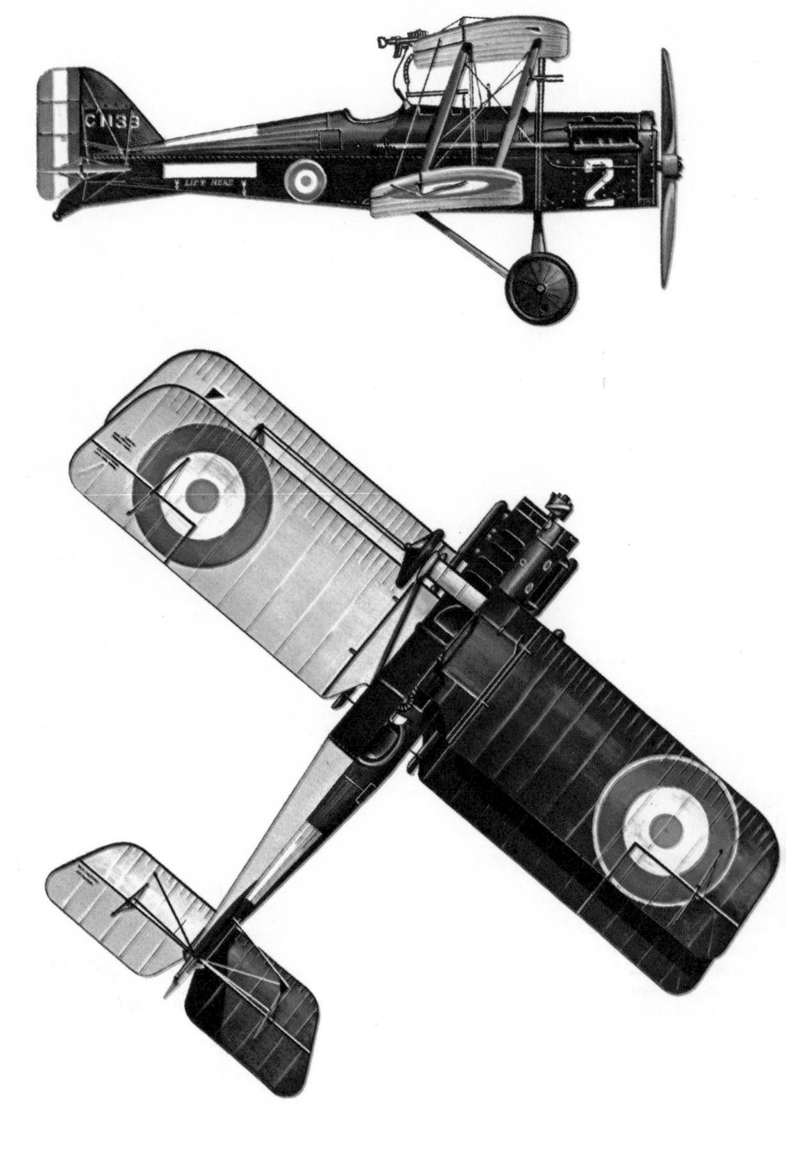

Royal Aircraft Factory-built SE5A flown by Capt K L Caldwell, commanding No 74 Sqdn, RFC, France, March 1918

Span: 26ft 7⅜in (8.11m)
Length: 20ft 11in (6.38m)
Weight: 1,988lb (902kg)
Engine: 200hp Wolseley W.4a Viper 8-cyl V-type
Max speed: 120mph (193.1kmh) at 15,000ft (4,570m)
Operational ceiling: 19,500ft (5,945m)
Endurance: 3hr
Armament: 1×0.303in Vickers and 1×0.303in Lewis machine-guns

The first of three Royal Aircraft Factory SE5 prototypes flew on 22 November 1916. It had a 150hp Hispano-Suiza, a 27ft 11in (8.51m) span and was one of the first aircraft to use Constantinesco synchronising gear for its forward-firing Vickers. Faster and stronger than Spads and Nieuports, but less manoeuvrable than the latter, it had excellent altitude performance, firepower and view from the cockpit. The first SE5s had cockpits semi-enclosed by 'canopies', the Vickers offset to port and a Lewis on a Foster mounting above the top wing centre-section. The next batch had a 26ft 7½in (8.12m) span. The first SE5s were delivered to No 56 Sqdn RFC in March 1917. Their 'canopies' were replaced by windscreens before their first operational sortie on 22 April in France. Nos 24, 60 and 85 Sqdns received some SE5s, which served until the end of 1917. The Factory built 59 SE5s with Hispano-Suiza 8As.

Meanwhile the third SE5 became the prototype SE5A, with a 200hp Hispano-Suiza and short-span wings. Deliveries began in June 1917 but were hampered by technical and supply trouble with engines. British-built alternatives were used, the best being the W.4a Viper, a re-designed, direct-drive Hispano-Suiza, which became the standard powerplant and gave it its 'square-jaw'. However, the SE5A acquired a fine reputation for its flying qualities, strength and performance, and shared with the Sopwith Camel chief credit for regaining Allied air superiority in 1918. The SE5A was used in the Middle East and by a few Home Defence units in 1918. It was less successful as a night fighter, being difficult to land at night, and its liquid-cooled engine taking longer to warm up than a rotary. At the Armistice, some 2,700 SE5/5As were on RAF charge, having served with 24 British, two US and one Australian squadrons. Serial allocations were made for nearly 5,500 SE5As; at least 200 were cancelled, and not all the others were delivered. In the USA, Curtiss assembled 56 from British components, but built only one all-American SE5A, flown on 20 August 1918, of a contract for 1,000 SE5As. The SE5B, a sesquiplane version, was not produced. Postwar, the SE5A went out of RAF service, but 50 were supplied to Australia, and others to Canada, South Africa, Poland and the USA; 50 came on to the British civil register, and became known in Britain chiefly for pioneerin in skywriting.

Bristol F2B Fighter

Bristol-built F2B (E2459) of B Flight, No 88 Sqdn, RAF, France, summer 1918

Span: 39ft 3in (11.96m)
Length: 25ft 10in (7.87m)
Weight: 2,779lb (1,261kg)
Engine: 275hp Rolls-Royce Falcon III 12-cyl V-type
Max speed: 123mph (198kmh) at 5,000ft (1,524m)
Operational ceiling: 18,000ft (5,485m)
Endurance: 3hr
Armament: 1×0.303in Lewis and 1× 0.303in Vickers machine-guns
Max bomb load: 240lb (109kg)

The Bristol Fighter was F S Barnwell's most successful design for the British and Colonial Aeroplane Co. Developed from the R2A/R2B 2-seat reconnaissance biplanes, the first prototype F2A with a 190hp Rolls-Royce Falcon flew on 9 September 1916 , the second (Hispano-Suiza) on 25 October. Fifty Falcon-engined F2As were built, with a forward synchronised Vickers gun and a Lewis in the rear cockpit. Deliveries began in December 1916, the first F2A squadron (No 48) arriving in France in March 1917. At first they were flown like previous 2-seaters, using the observer's gun as the primary weapon, and losses were heavy but when flown as a front-gun fighter, the Bristol achieved remarkable success. Many hundred F2Bs were ordered, with wider-span tailplanes, modified lower centre-sections and improved view. The first 150 F2Bs had Falcon Is, the next 50 220hp Falcon IIs and subsequent batches should have had the superb 275hp Falcon III. In July 1917 the War Office ordered 800 F2Bs to re-equip all RFC fighter reconnaissance and corps reconnaissance squadrons. The supply of Falcons could not keep pace and alternatives were sought, the 200hp Sunbeam Arab being most widely used. Re-equipment of squadrons in France with F2Bs quickened from summer 1917. By spring 1918 it had established such a reputation that enemy fighters would not attack more than two. The F2B served with six Western Front squadrons: Nos 11, 20, 22, 48, 62 and 88. In Palestine, F2Bs served with Nos 67 (Australian) and No 111 Sqdns. No 139 Sqdn in Italy and Home Defence Sqdns 33, 36, 39, 76 and 141 had F2Bs. Under wartime contracts over 5,250 F2A/Bs were ordered; 3,101 are known to have been accepted by RFC/RAF units by the Armistice. Plans for large-scale US production met disaster and only 27 of 2,000 ordered were built. Civil conversions appeared in 1919. After 1918 the RAF ordered a further 378. Bristol built 49 for export, delivering 137, including war surplus aircraft, to Belgium, Eire, Greece, Mexico, New Zealand, Norway and Spain. SABCA in Belgium built 40, and Curtiss 27 O-1 and Dayton-Wright 40 XB-1A US models. RAF F2Bs of 1919-20 were designated Mark II, having desert equipment and tropical radiators. The Mark III of 1926 incorporated structural redesign permitting higher operating weights. The designation Mark IV denoted marks modified with Handley-Page slots. Postwar, F2Bs served in the UK (Nos 2, 4, 13, 16 and 24 Sqdns), Egypt (47 and 208), Germany (5, 11 and 12), India (5, 20, 27, 28 and 31), Iraq (6 and 8), Palestine (No 14), and Turkey (4 and 208). The RAF retired its F2Bs in 1932, but the RNZAF's served until 1936.

Handley Page O/400

Span: 100ft 0in (30.48m)
Length: 62ft 10½in (19.16m)
Weight: 13,360lbs (6,060kg)
Engines: 2×360hp Rolls-Royce Eagle VIII 12-cyl V-type
Max speed: 97.5mph (156.9kmh) at sea level
Operational ceiling: 8,500ft (2,590m)
Endurance: 8hr
Armament: 3 to 5×0.303in in Lewis machine-guns
Max bomb load: 1,650lb (748kg) or 2,000lb (907kg)

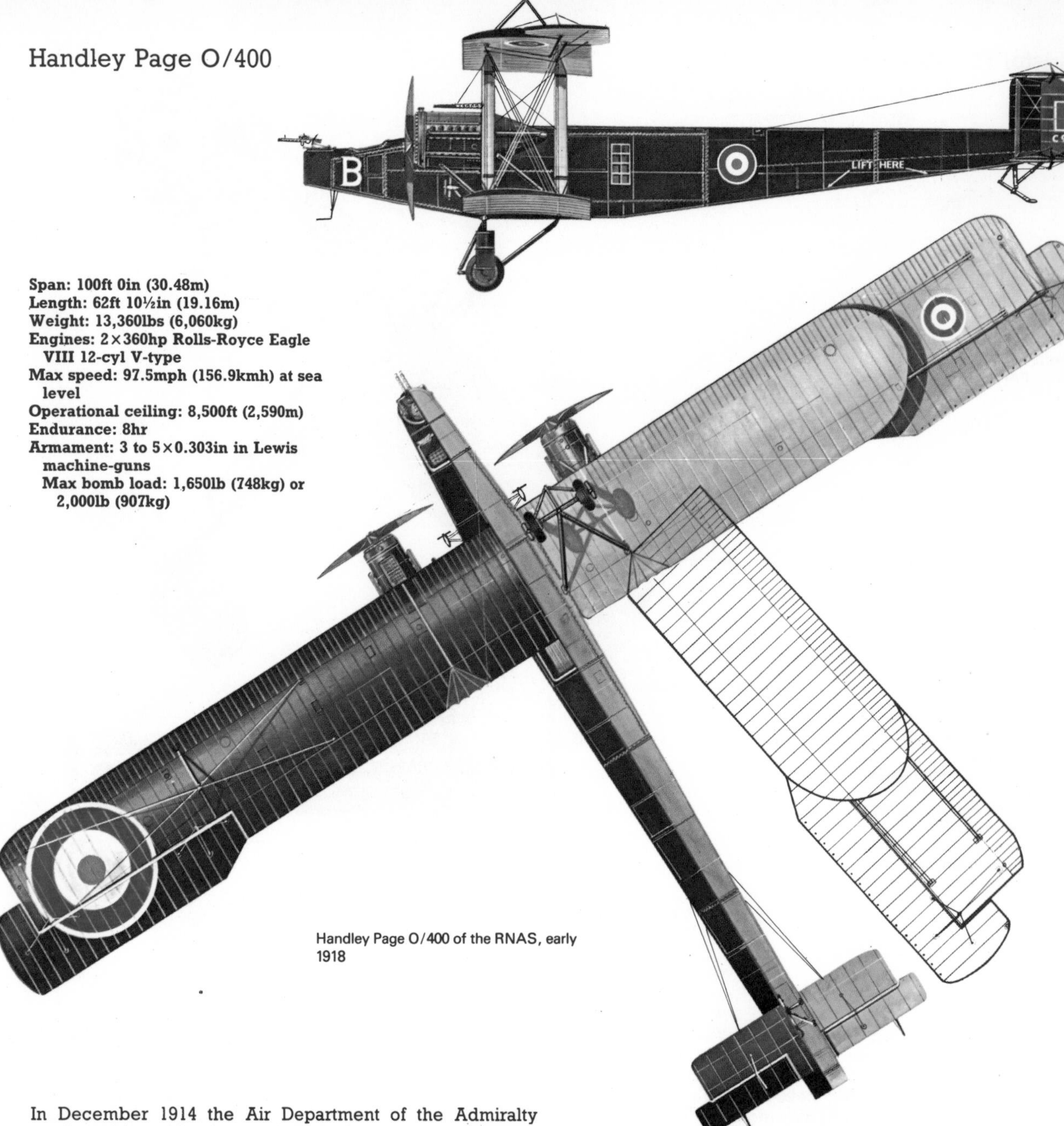

Handley Page O/400 of the RNAS, early 1918

In December 1914 the Air Department of the Admiralty confirmed in writing its requirement for a 'bloody paralyser of an aeroplane' for bombing Germany. By March 1915 40 O/100s had been ordered, the first flying on 18 December 1915, with two 250hp Rolls-Royce Eagle IIs, which powered 40 of the 46 O/100s built, the others having 320hp Sunbeam Cossacks. To fit field hangars, the wings folded back along the fuselage. It carried 16 112lb or eight 250lb bombs internally, and had a crew of four, and single or twin Lewis guns in nose and dorsal locations and a fifth ventral gun. The O/100 entered service with the RNAS on the Western Front in November 1916. Some O/100s served until the war ended. For a few months they were used for daylight sea patrols off Flanders, but from March/April 1917 concentrated on night bombing major German installations.

The much more numerous O/400 followed. The basic difference was that the fuel tanks were transferred from the engine nacelles to the fuselage, giving the O/400 much shorter nacelles. Successively higher-powered Eagle IV, VII or VIII engines were installed in O/400s, though some had the 260hp Fiat A-12bis or 275hp Sunbeam Maori. Nearly 800 O/400s were ordered during the war, of which about 550 were built; 107 were also assembled from components built in the USA with 350hp Liberty 12Ns. Eight of a 1,500 order were completed for the US Army. In April 1917 the O/400 became operational as a day bomber in France, transferring to night bombing from October. O/400s served with No 58 Sqdn RAF, Nos 97 and 115 Sqdns, Independent Force, and Nos 207, 214, 215 and 216 Naval Sqdns. A total of 258 O/400s were on RAF charge on 31 October 1918. One operated in Palestine. During the last months of the war the loads carried included the 1,650lb bomb. The O/400 remained in RAF service until 1920, eight being allocated as VIP transports for Paris Peace Conference officials in 1919. Four were used in 1919-20 for route-proving on Imperial Airways' air routes.

Sopwith F1 Camel

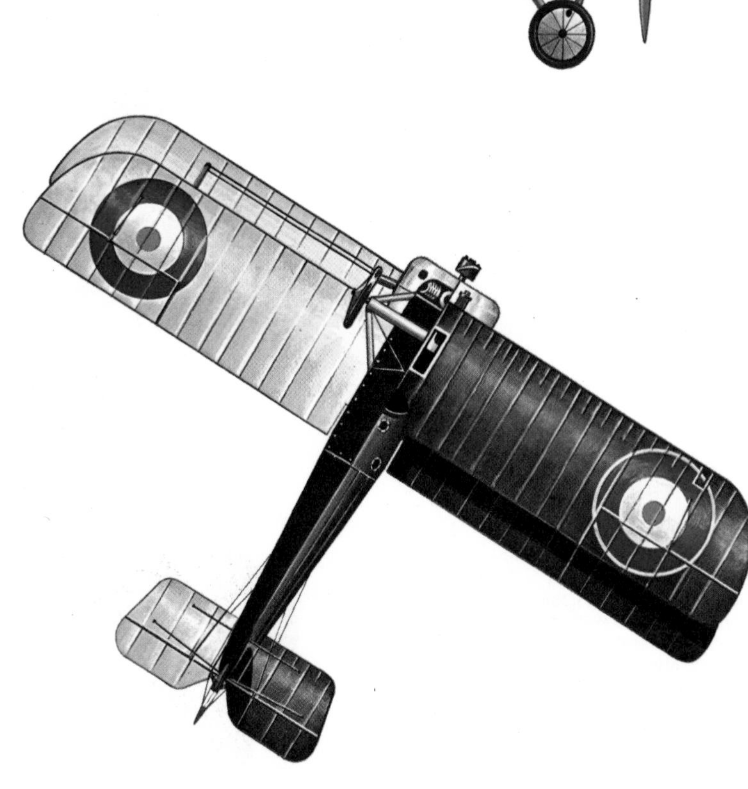

Boulton and Paul-built F1 Camel of No 65 Sqdn, RAF, France, mid-1918

Span: 28ft (8.53m)
Length: 18ft 9in (5.72m)
Weight: 1,453lb (659kg)
Engine: 130hp Cherget 9B 9-cyl rotary
Max speed: 104.5mph (168kmh) at 10,000 ft (3,048m)
Operational ceiling: 18,000ft (5,486m)
Endurance: 2hr 30min
Armament: 2×0.303in Vickers machine-guns
Max bomb load: 80lb (97kg)

The Camel, designed by Herbert Smith to succeed the Sopwith Pup and triplane, had none of the docile handling qualities of its predecessors, and had to be mastered before it could be flown successfully, but Camels were credited with destroying more enemy aircraft than any other Allied type. Aided by engine torque, the Camel could out-turn any German fighter, except possibly the Fokker Dr.I. Its name was originally unofficial, derived from the humped appearance created by the fairing over the Vickers gun breeches. The prototype appeared in December 1916, powered by a 110hp Clerget 9Z. Production Camels had either 130hp Clerget 9Bs or 150hp Bentley BR1s; deliveries began in May 1917. The first unit to receive Camels was No 4 Sqdn RNAS, with whom they became operational in July. Concurrent RFC orders specified either the Clerget 9B, BR1 or 110hp Le Rhône 9J engine. Camels had twin synchronised Vickers guns in front of the cockpit. Clerget Camels were faster; but Le Rhône Camels had a faster climb. By July 1917 first RFC deliveries had been made to No 70 Sqdn, and by the end of the year 1,325 Camels (of 3,450 on order) had been delivered. They were used widely for ground attack during the Ypres and Cambrai battles with four 20lb Cooper bombs under the fuselage, but losses among ground-attack Camels were heavy. A few Camels had 100hp Monosoupape engines. The USA purchased 143 Clerget-powered Camels in June 1918. These were later re-engined, detrimentally, with 150hp Monosoupapes, and issued to four AEF squadrons. From August 1917 the Le Rhône Camel was employed for Home Defence duties. All Camels so far discussed were F1 landplane fighters, but a shipboard version, the 2F1, underwent official trials in March 1917. Visually similar, it differed in construction from the F1, with a 1ft 1in (0.33m) shorter span, and the starboard Vickers deleted for an upward-firing Lewis. A total of 189 2F1 Camels were built, but did not become operational until spring 1918. Powerplants remained the Clerget 9B or the BR1, and some 2F1s acted as dive bombers with two 50lb bombs. By the end of the war 2F1 Camels had served aboard 5 aircraft carriers, 2 battleships and 26 cruisers of the RN. Discounting prototypes and cancelled contracts, 5,490 Camels were ordered, though delivery of all has not been established; and it is uncertain whether this figure includes the 2F1. On 31 October 1918 the RAF had 2,548 F1s on charge, over half of them with Clerget engines, and 129 2F1s. The Camel was quickly replaced by the Snipe in postwar RAF service, but served with the Belgian *Aviation Militaire*, the CAF, the Royal Hellenic Naval Air Service, the Polish Air Force and the US Navy; two came on to the British civil register.

Albatros D.V

Albatros D.Va of *Jasta 5,* Imperial German Military Aviation Service, early 1918

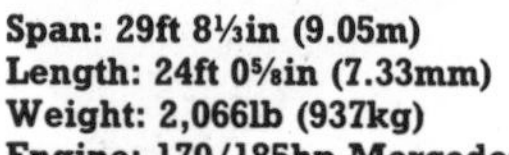

Span: 29ft 8⅓in (9.05m)
Length: 24ft 0⅝in (7.33mm)
Weight: 2,066lb (937kg)
Engine: 170/185hp Mercedes D.IIIa 6-cyl inline
Max speed: 116mph (187kmh) at 3,280ft (1,000m)
Operational ceiling: 20,505ft (6,250m)
Endurance: 2hr
Armament: 2×7.9mm Spandau machine-guns

The Albatros D.V was produced in a not entirely successful attempt to maintain the superiority gained in 1917 by the excellent D.III, in the face of later Allied types such as the SE5A and Sopwith Camel. An interim model, the D.IV, appeared in 1917, marking a return to the equal-chord wings of the D.II and powered by a fully enclosed geared Mercedes engine. Owing to troubles with this engine the D.IV was not developed, but its fuselage design was retained in the D.V, which resumed the more graceful and more efficient wing form of the D.III. The D.V's fuselage was of oval section (compared with the flat-sided D.III), and a high-compression Mercedes D.IIIa was installed with fewer pretensions to careful cowling to simplify access and maintenance. The D.V had the same tailplane as the D.III, but introduced an integral fixed fin, a raked-back underfin and a more rounded rudder. Fuselage construction was lighter but stronger than that of the earlier Albatros fighters, although the D.Va's gross weight was slightly increased over that of the D.V by additional strengthening. The D.V also differed from the other Albatros fighters in the arrangement of its aileron control wires; this was the only visible distinction between the D.V and D.Va. Unfortunately, although flying qualities remained good, the D.V and D.Va were not a great improvement over the D.III, achieving their success as much by their numbers as by performance. The D.I from mid-1916, then the D.II, and the D.III from early 1917 had wrested back air superiority from the French and British; the first D.Vs were delivered to *Jastas* in June 1917, the D.Vs following from late autumn, both versions serving alongside the earlier D.IIIs. They reached their peak of service in November 1917 and March 1918 respectively, and were the most widely used of all Albatros fighters. Exact production figures are not known, but a minimum of 1,512 D.V/Vas are known to have served with Western Front units, and this takes no account of aircraft with home establishments or those used in Italy and Palestine. Production was shared by the Ostdeutsche Albatros Werke. Despite limitations on diving manoeuvres, imposed after a series of crashes caused by failure of the single-spar lower wings (also a weakness of the D.III), D.V/Vas remained in service until the Armistice. In the post-war redesignation of Albatros types the D.V/Va became known as the L24.

Fokker Dr.I

Fokker Dr. I (474) believed to be an aircraft of *Jasta 2,* Imperial German Military Aviation Service, early 1918

Span: 23ft 7⅝in (7.19m)
Length: 18ft 11⅛in (5.77m)
Weight: 1,290lb (585kg)
Engine: 110hp Oberursel UR.II 7-cyl rotary
Max speed: 102.5mph (165kmh) at 13,125ft (4,000m)
Operational ceiling: 20,000ft (6,100m)
Endurance: 1hr 30min
Armament: 2×7.9mm Spandau machine-guns

Such was the impact created in German military circles by the Sopwith triplane that 14 German and Austrian manufacturers produced triplane designs. Most of them did so after inspecting a captured machine in July 1917, but they were well behind Anthony Fokker, who had seen the Sopwith in action at the Front in April. It has often been implied that the Fokker Dr.I *(Dreidecker* = triplane) was a copy of the Sopwith triplane, but Reinhold Platz, who designed the Fokker machine at his employer's request, had never seen the British aircraft, and indeed was unconvinced of the merits of a triplane layout. Nevertheless, he produced a prototype known as the V.3 with three sets of cantilever wings, the only struts being those on which the top plane was mounted. Single, thin interplane struts and balanced ailerons, and 2 7.92mm Spandaus, were added after Fokker test-flew the V.3 which in this form became the V.4. In virtually unchanged form, 2 further prototypes and 318 Dr.Is were ordered in summer 1917. Following acceptance trials in August, the second and third prototypes became operational with Manfred von Richthofen's *JGI.* The latter was flown almost exclusively by Lt Werner Voss, who scored his first victory with it on 30 August. In the next 24 days he scored a further 20 victories before being shot down and killed on 23 September by SE5As of No 56 Sqdn RFC. From mid-October, deliveries of production Dr.Is began to *JGI,* but by early November a series of fatal crashes caused them to be grounded. The trouble was faulty wing construction. The Dr.I thus did not become fully operational until late November, and its subsequent career at the Front was brief. Nevertheless, it achieved considerable success, due mainly to its excellent manoeuvrability and rate of climb that put it on equal terms with the SE5A, Bristol F2B and Spad. It was in a Fokker Dr.I that von Richthofen, its greatest exponent, was shot down and killed on 21 April 1918. The Dr.I reached its peak of service early in May 1918, when 171 were in front-line service. Later that month production ceased, and thereafter the Dr.I was transferred to home defence. At the Armistice, 69 were in service. The standard powerplant of the Dr.I was the 110hp Oberursel UR.II or Goebel Goe II rotary. Some had the 200hp Goe III or IIIa engine or the 110hp Le Rhône, captured or copied; experimental installations included the Sh. III and captured 130hp Clergets.

Gotha G.V

Span: 77ft 9⅛in (23.70m)
Length: 40ft 6¼in (12.35m)
Weight: 8,763lb (3,975kg)
Engine: 2×260hp Mercedes D.IVa 6-cyl inlines
Max speed: 87mph (140kmh) at sea level
Operational ceiling: 21,325ft (6,500m)
Range: 522 miles (840km)
Armament: 2×Parabellum machine-guns
Max bomb load: 1,100lb (500kg); normal: 660lb (300kg)

Gotha G.V of the Imperial German Military Aviation Service, April 1918

The first Gothaer Waggonfabrik AG *Grossflugzeug* (large aeroplane), the G.I, evolved from a prototype flown in January 1915. Gotha licence-built a few for ground attack and general tactical duties on the Western and Eastern Fronts. The G.Is had a slim fuselage attached to the upper wings; two 160hp Mercedes D.IIIs were mounted on the lower wing. Following the concept, the G.II was an entirely new design, flown in March 1916. The fuselage and engines (220hp Mercedes D.IVs) were mounted on the lower wings; span was increased and auxiliary front wheels prevented nosing over. It carried three crewmen, and two machine-guns. Entering service in autumn 1916, it was soon withdrawn after repeated engine failures. It was replaced from October 1916 on the Balkan and Western Fronts by the G.III, with reinforced fuselage, three machine-guns and 260hp Mercedes D.IVas. An initial 25 G.IIIs were ordered, and in December 1916 14 were at the Front. First major model was the G.IV (Mercedes D.IVa), chosen for raids on the UK: an initial 50 were ordered from Gotha; 80 were built by Siemens-Schuckert and some 100 by LVG. It began daylight raids on southern England in late May 1917. With an all-plywood fuselage, and ailerons on both wings, it was stronger yet easier to fly than its predecessors, though, as performance remained similar to the G.III's, it had to be switched to night attacks against Britain from September. It was replaced by the G.V from August, which continued the night bombing of England until May. At peak employment, in April 1918, 36 G.Vs were in service. Typical bomb load on cross-Channel raids was six 50kg bombs. Final versions in service were the G.Va/Vb, differing in internal details, but distinguishable from the G.V by their biplane tail assembly and shorter nose. They went into production in March 1918 and into service in June; by August there were 21 at the Front. They were agile for their size, well defended and difficult to shoot down. More were lost to AA fire than in aerial fighting but far more were lost in landing accidents. Siemens built 40 G.IVs as trainers, most with 180hp Argus As.IIIs or 185hp NAGs. About 30 LVG G.IVs were later transferred to Austro-Hungary and employed on the Italian Front, with 230hp Hieros.

Caproni Ca 5

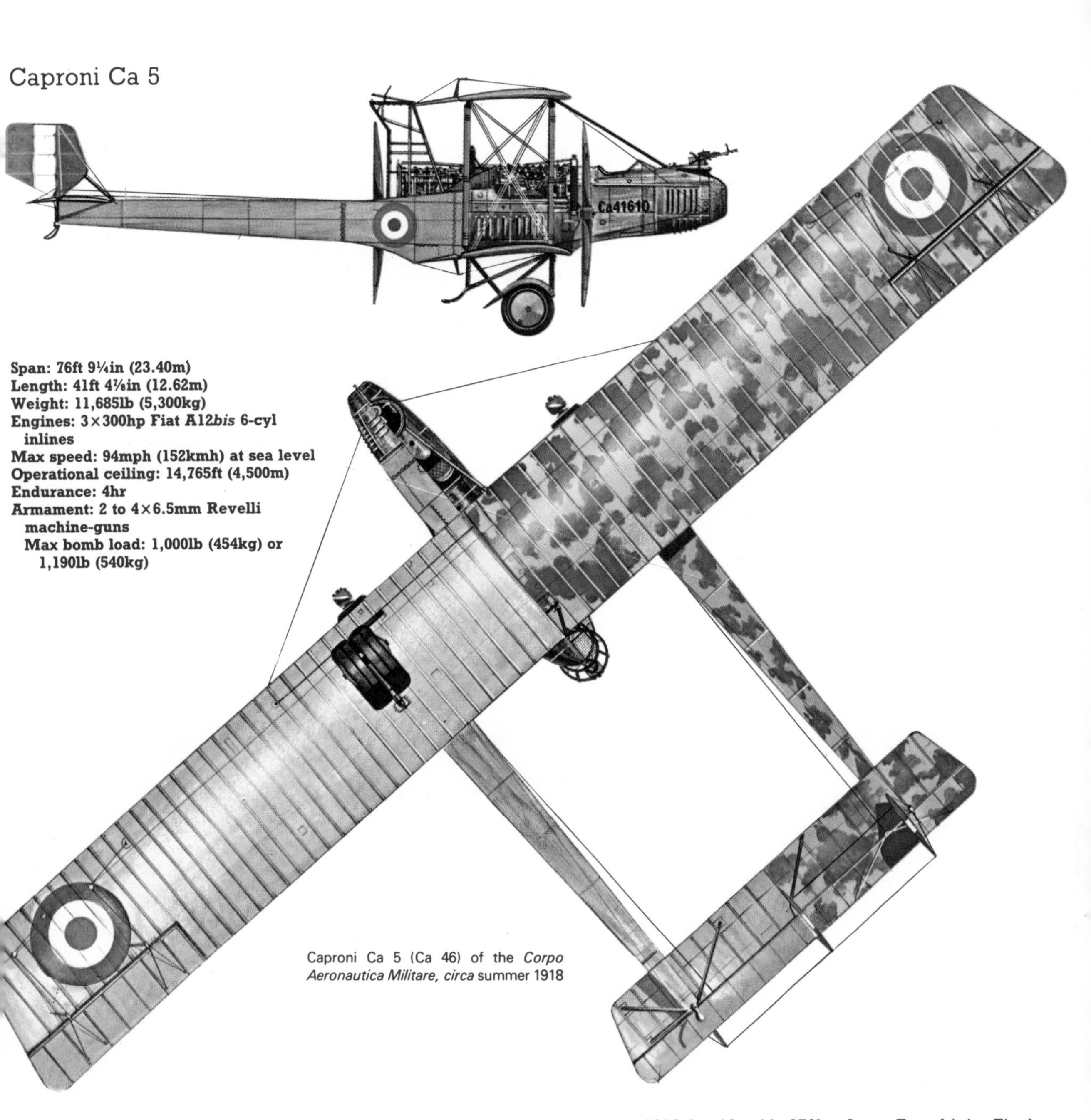

Span: 76ft 9¼in (23.40m)
Length: 41ft 4⅜in (12.62m)
Weight: 11,685lb (5,300kg)
Engines: 3×300hp Fiat A12*bis* 6-cyl inlines
Max speed: 94mph (152kmh) at sea level
Operational ceiling: 14,765ft (4,500m)
Endurance: 4hr
Armament: 2 to 4×6.5mm Revelli machine-guns
Max bomb load: 1,000lb (454kg) or 1,190lb (540kg)

Caproni Ca 5 (Ca 46) of the *Corpo Aeronautica Militare, circa* summer 1918

Italy and Russia had evolved aircraft suitable for heavy bombing well before the outbreak of World War I. The first Caproni giant was designed in 1913, setting the pattern for developments. It had three-bay, equal-span wings, a central nacelle and two slender fuselage booms supporting three polygonal rudders atop the tailpanes. The prototype flew in October 1914. The first production version, designated Ca 1 (162 built), had three 100hp Fiat A-10 inlines driving tractor propellers. Ca 1s made the first Italian bombing raids of the war against Austro-Hungary on 20 August 1915 (before the HP O/100 had flown). Nine, re-designated Ca 2, had the central engine replaced by a 150hp Isotta-Fraschini V-4B inline. The Ca 1 and 2 served, latterly on night operations, until the Ca 3 appeared in 1917. Similar to the Ca 1, the Ca 3 had three 150hp Isotta-Fraschini V-4Bs. Several were still in service at the Armistice. The military designation Ca 4 applied to a series of much larger triplanes. The first three, appearing in late 1917 had three 200hp Isotta-Fraschinis, followed in 1918 by 12 with 270hp Isotta-Fraschinis. Final production variants (23 built) had more powerful Fiat, Isotta-Fraschini or Liberty engines. Too slow for daylight bombing, the Ca 4 was employed principally at night. The Ca 5 returned to the biplane configuration, and was slightly bigger than the Ca 3. First model in the series, which entered service in early 1918, had three 200hp Fiats. The wing-mounted pair had frontal radiators while that for the middle engine was in the nacelle nose. Improved versions followed, with three 200hp Isotta-Fraschini, Fiat of Liberty engines. The Ca 5 operated mostly with Italian night-bomber squadrons in France, although some were still flying on the Italian Front at the Armistice. Total Italian production of the Ca 5 reached 250; a small batch were built by Esnault-Pelterie in France. Two Ca 5s were delivered to the USA, and three were built there before the war ended. Their load may not seem exceptional but their targets were usually at great distance, involving flights over mountain country.

Fokker D.VII

Fokker D.VII (510), possibly Albatros-built, believed to be an aircraft of *Jasta 17*, Imperial German Military Aviation Service, *circa* May 1918

Span: 29ft 2⅓in (8.90m)
Length: 22ft 9¾in (6.954m)
Weight: 2,116lb (960kg)
Engine: 160hp Mercedes D.III 6-cyl inline
Max speed: 117.4mph (189kmh) at 3,280ft (1,000m)
Operational ceiling: 19,685ft (6,000m)
Endurance: 1hr 30min
Armament: 2×7.9mm Spandau machine-guns

The Fokker D.VII, widely claimed as the best German fighter of World War I, was developed to a specification of late 1917. Its true prototype was the Fokker V.11, designed by Reinhold Platz. The V.11 was tested at Adlershof in January/February 1918. It was superior to the other 30 entrants by a wide margin, and with modifications made at Manfred von Richthofen's instigation was immediately ordered for large-scale production: 400 from Fokker, and substantial quantities from Albatros and OAW. The V.11 was unstable in a dive, and production D.VIIs therefore had a lengthened fuselage and a fixed fin. The view from the cockpit was excellent and the D.VII was armed with two 7.92mm Spandau guns, with 500rpg. It was easy to fly, but its main advance over earlier German fighters was its performance at high altitude, enhanced from late summer 1918 by the D.VIIF, powered by a 185hp BMW IIIa. The D.VIIF had greater power reserves above 5,000m (16,400ft), which height the D.VIIF could reach in 14 minutes, compared with the D.VII's time of 38 minutes. Von Richthofen's *JGI* (later commanded by Hermann Goering) received the first Fokker D.VIIs in April 1918. The custom was to allocate new fighters to *Jastas* and pilots in order of eminence, and several months elapsed before lesser *Staffeln* received D.VIIs. Nevertheless, by the Armistice 760 had been accepted and were operated by 48 *Jastas,* although several were below establishment. Fokker built at least 840 D.VIIs; 785 were ordered from Albatros, and 975 from OAW. In Austro-Hungary it was built by MAG as the Series 93. The Fokker D.VII was respected like no German fighter since the Fokker E.III. The Armistice Agreement 'honoured' it by specifically mentioning it among equipment to be handed over to the Allies, squashing Anthony Fokker's hopes of continuing aircraft manufacturing in Germany after the war. However, he succeeded in smuggling 400 engines and components of 120 aircraft, most of them D.VIIs, out of Germany into Holland where the D.VII continued in production after the war, and remained in service, with the Dutch Army Air Service and later in the Netherlands East Indies, until the late 1920s. Between 1919 and 1926 ex-wartime D.VIIs were used, after conversion to 2-seaters, as trainers by the Belgian *Aviation Militaire;* 27 were supplied to the Swiss *Fliegertruppe.*

Halberstadt C.V

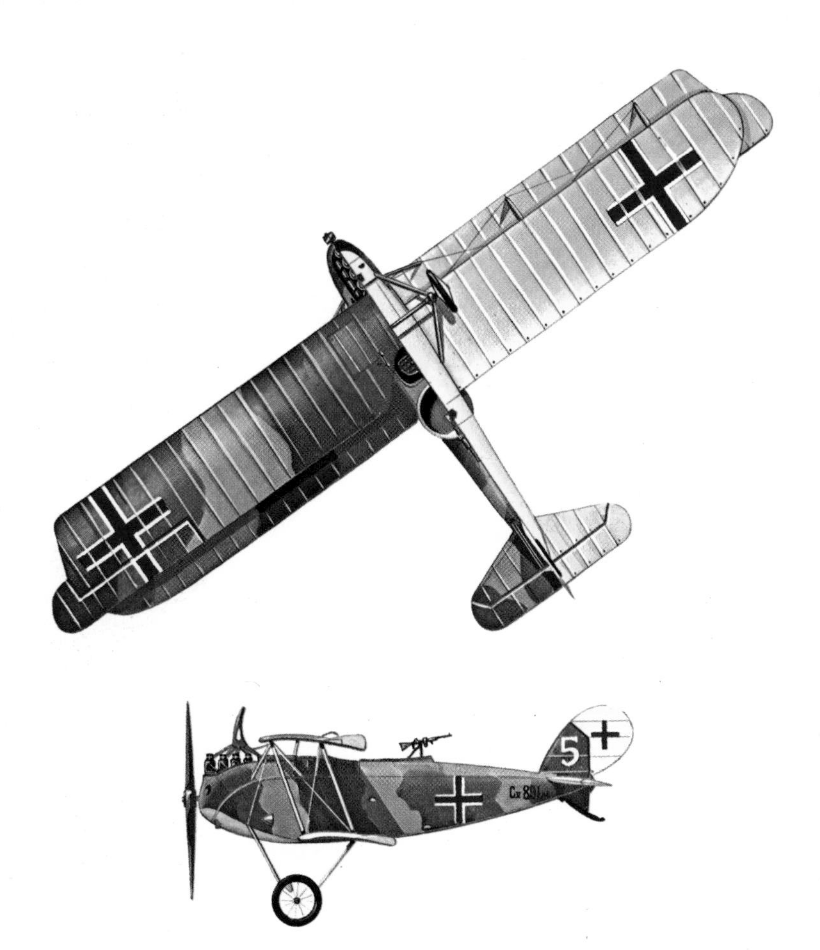

Halberstadt C.V (possibly Aviatik-built) of the Imperial German Military Aviation Service, *circa* summer 1918

Span: 44ft 8⅓in (13.62m)
Length: 22ft 8½in (6.92m)
Weight: 3,009lb (1,365kg)
Engine: 220hp Benz Bz.IV 6-cyl inline
Max speed: 105.6mph (170kmh) at sea level
Operational ceiling: 16,405ft (5,000m)
Endurance: 3hr 30min
Armament: 1×7.9mm Spandau, and 1×6.5mm Parabellum machine-guns
Max bomb load: 110lb (50kg)

The two principal types employed by the German air force in 1918 for photographic reconnaissance work were the excellent Rumpler C.VII and the Halberstadt C.V. The Halberstadt company's previous C types began with the C.I, a 1916 adaptation, with a rotary engine, of the unarmed B.II. It is doubtful whether this went into production. The C.III, which appeared late in 1917, was the first long-range photographic type designed by Karl Theiss. It was powered by a 200hp Benz Bz.IV and an unusual feature of its design was the attachment of the lower wings to a small 'keel' on the underside of the fuselage. The C.III formed the basis for the C.V, developed in late 1917 by Theiss, in which a simpler and more conventional attachment was employed for the lower wings. Powerplant of the C.V was the high-compression version of the Bz.IV developing 220hp and giving a much better performance at altitude than that of the C.III. The wings were of wide span, with two bays of bracing struts and overhung, balanced ailerons on the upper surfaces. The fuselage was essentially a scaled-up version of that used in the C.IV, but with separate cockpits for the 2-man crew. Reconnaissance cameras were aimed downward through a trap in the floor of the rear cockpit; the top of the cockpit was built up with the traditional ring mounting for a Parabellum machine-gun at the rear. The pilot was furnished with a synchronised 7.92mm Spandau machine-gun immediately in front of his cockpit on the port side; C.Vs also normally carried wireless telegraphy equipment. The prototype C.V appeared early in 1918, undergoing its official trials in the spring. Its front-line career lasted from summer 1918 until the Armistice. Production was undertaken by the Aviatik, BFW and DFW companies in addition to those built by Halberstadt. Further Halberstadt variants appeared during 1918 including the C.VII, C.VIII and C.IX. The C.VII (245hp Maybach Mb.IV) and C.IX (230hp Hiero) remained in the prototype stage, though the latter may have been intended for Austro-Hungarian production. The C.VIII, officially tested in October 1918, was a single-bay biplane, slightly smaller than the C.V. and powered by an Mb.IV engine; it had a ceiling of 9,000m (29,528ft), which it could reach in 58 minutes, and was probably intended for series production if the war had continued; however, only the prototype had been completed when the war ended.

Breguet XIX

Breguet XIXB2 of an Army Group of the Greek National War Aviation Forces, *circa* 1930

Span: 48ft 7¾in (14.83m)
Length: 31ft 2in (9.50m)
Weight: 4,850lb (2,200kg)
Engine: 500hp Hispano-Suiza 12Hb 12-cyl V-type
Max speed: 143mph (230kmh) at 9,845ft (3,000m)
Operational ceiling: 21,980ft (6,700m)
Range: 497 miles (800km)
Armament: 3 or 4×7.5mm Darne (or 7.7mm Vickers) machine-guns
Max bomb load: 970lb (440kg)

Designed to replace the successful Breguet Type XIV World War I bomber, the Breguet Type XIX was designed in 1921 and made its first public appearance as an exhibit at the *Salon de l'Aéronautique* in Paris that year. At that time it had not yet flown; it made its maiden flight in May 1922, powered by a 420hp Renault engine, which was later exchanged for a 375hp and then for a 450hp Lorraine V-type engine. The Bre XIX was of metal construction, with fabric covering, and its clean lines and single-strut sesquiplane layout made it straightforward to build and maintain. A feature of the design was the ease with which a large variety of powerplants could be fitted. During its 15-year operational life the Bre XIX was powered by 400 and 450hp Lorraine, 420 or 480hp Jupiter, 480, 500 or 550hp Renault, 450 or 500hp Hispano-Suiza, 500hp Farman and 500hp Salmson engines. A twin-float landing gear could be substituted for the more usual landplane undercarriage. French production, which began in 1925, reached about 1,100 examples, and at its peak was turning out four Bre XIXs a day. Despite a multiplicity of variants, created by the many permutations of powerplant, landing gear, armament and detail features, there were only three basic functional versions of the aircraft. These were the Bre XIXA2 (2-seat observation/reconnaissance), the Bre XIXB2 (2-seat day and night bomber) and the Bre XIXGR *(grande reconnaissance).* All three served with the French *Aviation Militaire,* and substantial exports of A2 and B2 models were made to Argentina, Belgium, Bolivia, China, Greece, Persia, Poland, Romania and Yugoslavia. The Bre XIX was also built under licence in Belgium (approximately 150 by SABCA), Greece, Japan, Spain (103 by CASA) and by Ikarus in Yugoslavia. The Breguets began to disappear from first-line French bomber squadrons in the latter 1930s, although 116 were still on charge at 1 January 1936. Some foreign users kept them in service until at least 1940. Individual Bre XIXs were responsible for endurance and other record flights during the 1920s and 1930s, perhaps the most noteworthy being the *Point d'Interrogation* (Question Mark), flown by the Frenchmen Dieudonné, Costes and Maurice Bellonte on the first successful east-to-west heavier-than-air flight across the North Atlantic, from Paris to New York, on 1-2 September 1930. Another famous Bre XIX, the *Nungesser et Coli,* is in the possession of the Musée de l'Air at Le Bourget.

Nieuport-Delage 62

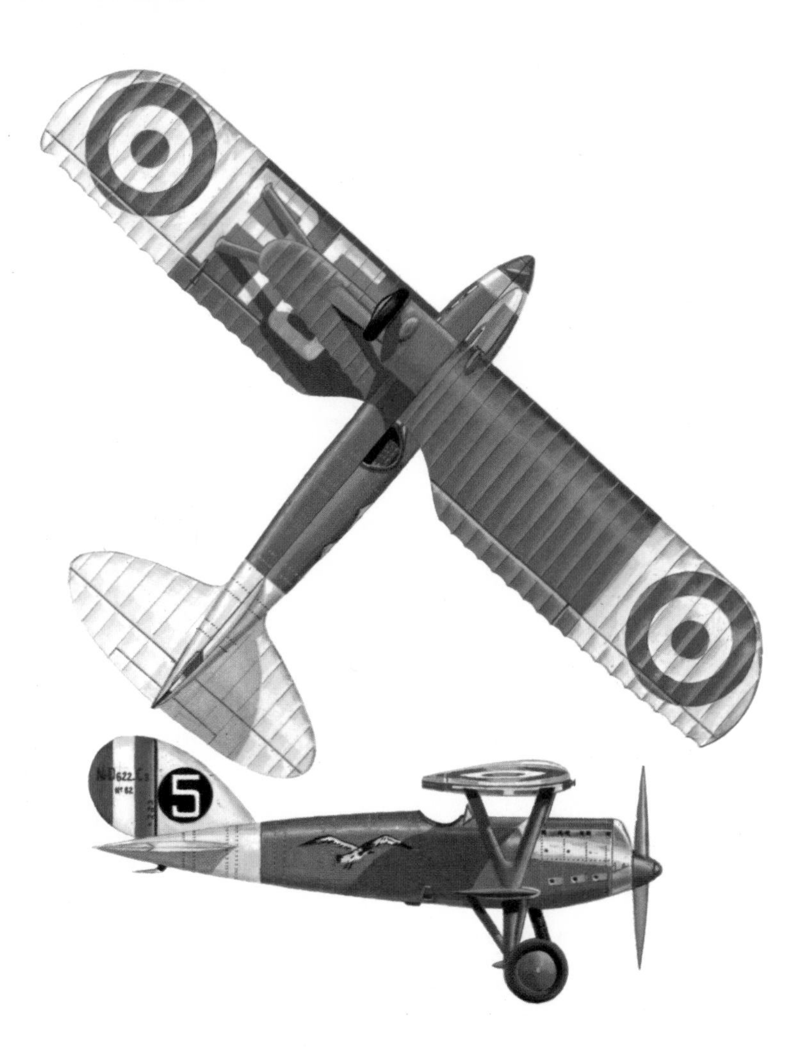

Nieuport-Delage 622C.1 of the *4e Escadrille, 2e Escadre de Chasse, Armée de l'Air,* in markings for 'Defence of Paris' exercises, 1934

Span: 39ft 4½in (12.00m)
Length: 25ft 0¾in (7.64m)
Engine: 500hp Hispano-Suiza 12Md 12-cyl V-type
Max speed: 168mph (270kmh) at sea level
Operational ceiling: 26,905ft (8,200m)
Range: 311miles (500km)
Armament: 2×7.7mm machine-guns

This line of French sesquiplane fighters spanned 15 of the years between the world wars, the NiD 62 series being one of the three major biplane types in service during the 1930s until and after the advent of the Dewoitine 500/510 monoplanes. Origin of the species lay in the NiD 42S, a racing aircraft of 1923 from which was developed in 1924 the single-seat NiD 42 high altitude fighter. It was designed in two alternative configurations – parasol monoplane and sesqiplane – and, in 1925, 25 of the later were ordered for the French Air Force as the NiD 42C1. Various versions of the NiD 42 were later flown but no further series production was undertaken. In 1928 the NiD 52 appeared, with a metal-clad fuselage. Dimensionally identical to its predecessor, and likewise powered by a 500hp Hispano-Suiza HS 12Hb liquid-cooled engine, it was 15mph (24kmh) slower, depite little increase in gross weight. It was produced by Hispano in Spain, who built 125, of which 36 were in service when the Civil War broke out in 1936; the capture of some by the Nationalist forces led to combats between opposing pilots using the same aircraft. The major type in this sequence, however, was the NiD 62 of 1928, whose principal outward differences from the NiD 52 were broader-chord wings with shorter ailerons, an enlarged tailplane and an HS 12Mb engine and wooden propeller. A total of 345 was built for the French Air Force. The NiD 622 (330 built) followed, with HS 12Md supercharged engines, metal propellers and ailerons over the trailing edge outboard of the centre-section cut-out. Twelve aircraft, with Lorraine 12Hdr engines but otherwise similar, were built in 1933 for Peru; three fabric-fuselage NiD 62s (redesignated NiD 72) were supplied to Belgium in 1929. Final version produced for the French Air Force was the NiD 629 (50 built), with a different supercharger and oleo-type landing gear. The line ended with a solitary NiD 82, originally a Lorraine-engined all-metal sesquiplane with square wingtips. It appeared in 1931, but was converted later in the year to a monoplane and sold to Spain. By the end of 1932, the 33 first-line fighter squadrons in metropolitan France possessed 366 aircraft. Of these, 251 were NiD 62/622s, compared with 70 LGL-32s and 45 Wibault 72s. Three years later, following the introduction of the Dewoitine 500 series, NiD 622s equipped only one *groupe* of the *5e Escadre* at Lyon-Bron. But many had been transferred to *Escadrilles Regionales,* with whom some were still operational in September 1939; and during May 1940, when anything flyable was used against the Germans, *GC*s III/4 and III/5 flew a number of NiD 629s.

Bristol Bulldog

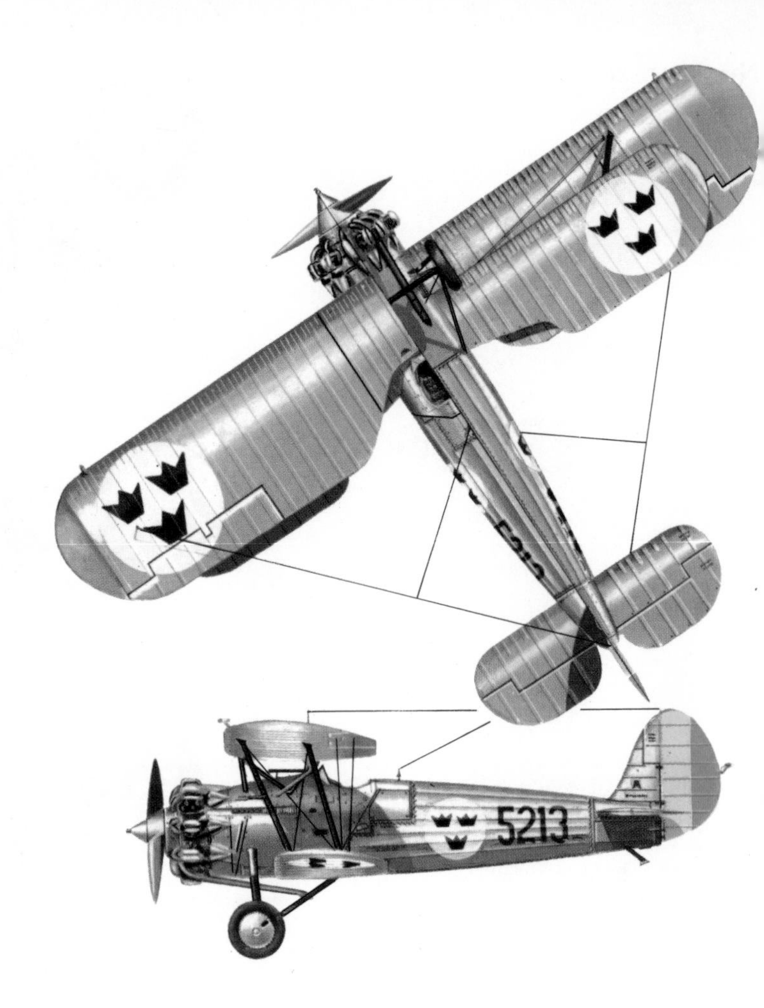

Bristol Bulldog Mk IIA delivered to the Royal Swedish Air Force, 1931

Span: 33ft 10in (10.31m)
Length: 25ft 2in (7.67m)
Weight: 3,530lb (1,601kg)
Engine: 490hp Bristol Jupiter VIIF 9-cyl radial
Max speed: 178mph (286kmh) at 10,000ft (3,050m)
Operational ceiling: 29,300ft (8,930m)
Range: approx 275 miles (443km)
Armament: 2×0.303in Vickers machine-guns
Max bomb load: 80lbs (36kg)

The Bulldog was of the 'second generation' of post-war RAF fighters, reflecting a change in both strategic and performance requirements. One of several excellent fighter designs by Captain F S Barnwell of the Bristol Aeroplane Company, it was submitted in 1927 (with optional Rolls-Royce Falcon inline or Bristol Mercury radial) to meet an RAF requirement for an interceptor fighter to replace the Armstrong Whitworth Siskin. Two prototypes were built, one to official order (Bristol Type 107 Bullpup J9051, Mercury III engine) and one as a private venture (Type 105 Bulldog, Jupiter VII engine); the latter was first to fly, on 17 May 1927. Construction was all-metal, with fabric covering for all areas except the front part of the fuselage. A second Bulldog prototype (J9480) was ordered in November 1927; this had a lengthened rear fuselage and was designated Mk II. Service trials of the Mk II occupied the early months of 1928, being followed by an initial order for 25 production aircraft with Jupiter VII engines. Eventually, 92 Mk IIs were built, but the major service variant was the Bulldog Mk IIA, 268 of which were completed. These incorporated constructional improvements to permit a higher operating weight, and were powered by Jupiter VIIF engines. The first Mk IIs were delivered to Nos 3 and 17 Sqdns in May and October 1929, replacing Gamecocks and Siskins. Other RAF squadrons in the UK to have the Bulldog Mks II or IIA included Nos 19, 23, 29, 32, 41, 54, 56, and 111. They were employed for both day and night fighting duties, and by 1932 Bulldog squadrons represented over two-thirds of the total fighter defence of the UK. The Bulldog TM was a 2-seat trainer counterpart of the Mk IIA, with slightly-swept wings and an enlarged rudder; 59 were built. The final main version (18 were built) was the Mk IVA, which had a 46mph (74km/hr) faster speed than the Mk IIA thanks to its Mercury VIS2; 17 of these were to a Finnish Air Force order, several taking part in the 1939-40 Russo-Finnish winter war. Other foreign purchasers of Mks II/IIAs included Australia (8); Denmark (4); Estonia (12); Latvia (12); Siam (2); and Sweden (11). Three of the Swedish Bulldogs (Swedish designation J 7) were transferred to the Finnish Air Force for training in 1939. Two Bulldogs were completed by Nakajima, but no series production in Japan was undertaken. As an RAF fighter, the Bulldog remained in service until the advent of the Gladiator in 1937.

Boeing F4B

Boeing F4B-3 of VF-1B, US Navy, USS *Saratoga, circa* November 1932

Span: 30ft 0in (9.14m)
Length: 20ft 4¾in (6.22m)
Weight: 2,918lb (1,324kg)
Engine: Pratt & Whitney R-1340D Wasp 9-cyl radial
Max speed: 187mph (301kmh) at 6,000ft (1,809m)
Operational ceiling: 27,500ft (8,380m)
Range: 585 miles (941km)
Armament: 2×0.30in (or 1×0.30in and 1×0.50in) machine-guns
Max bomb load: 232lb (106kg)

The Boeing F4B and P-12 single-seat fighters were, respectively, US Navy and Army variants of the same basic design, the former being the first to be ordered into production. The fighter originated with the private-venture Boeing Model 83 and Model 89, of which the former flew for the first time on 25 June 1928, powered by a 400hp Pratt & Whitney Wasp radial engine. This aircraft, and the Model 89, were delivered to the US Navy some two months later for evaluation, and were given the designation XF4B-1. Successful trials were followed by an initial contract for 27 production aircraft, given the Boeing Model number 99 and the service designation F4B-1. The first F4B-1 was flown on 6 May 1929, and deliveries to the US Navy began in the following month. The US Army Air Corps was so impressed by the Navy's opinion of the Boeing fighter that it took the rare course of placing an order of its own, albeit a small one, without conducting an independent evaluation of the aircraft. Nine were ordered as P-12s (Boeing Model 102), plus a tenth machine designated XP-12A with Frise-type ailerons, modified landing gear and improved long-chord engine cowling. Subsequent Navy models were the F4B-2 (Frise ailerons, Townend cowling ring and divided-axle landing gear: 42 built); F4B-3 (similar, but with an all-metal fuselage: 21 built); and F4B-4 (enlarged headrest and broader fin and rudder: 74 built). Those ordered by the USAAC continued with the P-12B (90 built, with Frise ailerons and shortened landing gear); P-12C (96 with Townend rings and further undercarriage modifications); 35 P-12Ds and 110 metal-fuselage P-12Es with uprated R-1340-17 Wasp engines; and 25 P-12Fs with R-1340-19 radials. The Brazilian government purchased 23 of these fighters, 14 of which were standard Navy-type F4B-4s. The remaining 9, designated Boeing Model 267, were of hybrid type combining various features of the F4B-3 and P-12E. Two other examples (Model 100E) were sold to the Spanish government. Altogether, including test aircraft, a total of 586 fighters in this basic design series were completed by Boeing, a production total that was not exceeded by any other US warplane until the eve of America's entry into World War 2.

Hawker Hart

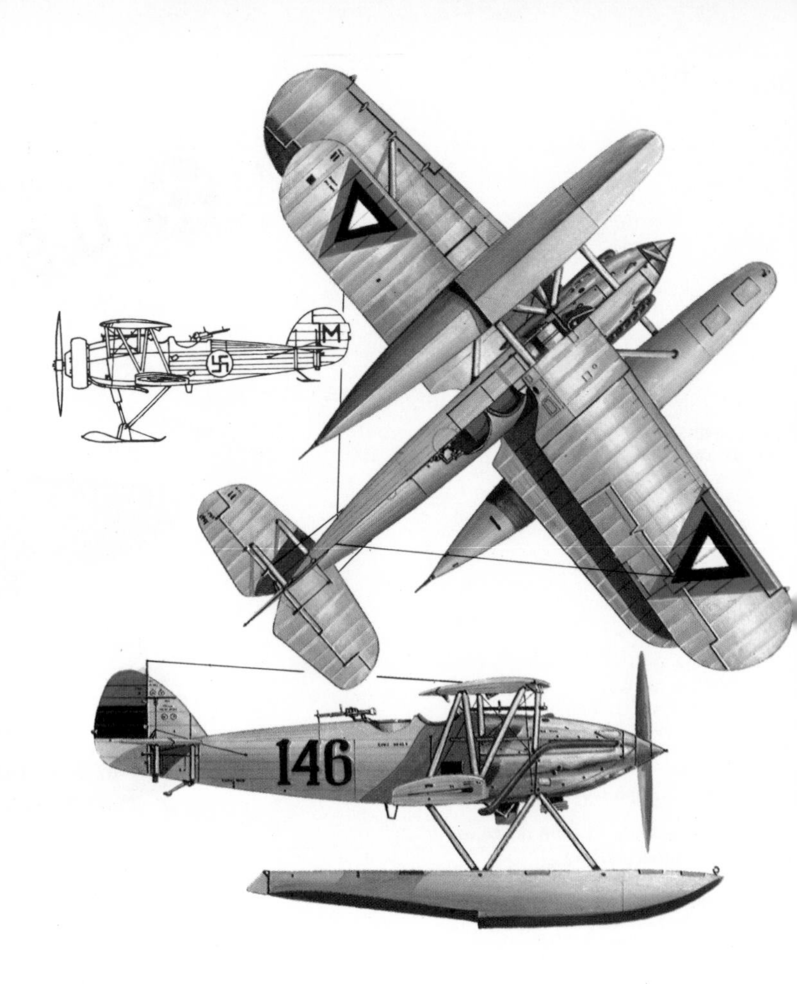

Hawker Hart floatplane of the Estonian Air Force, *circa* 1932. (Inset view shows version in service with Royal Swedish Air Force, with 580hp Pegasus IM2 radial engine and ski-undercarriage; weight and performance data are for landplane Hart.)

Span: 37ft 3in (11.35m)
Length: 29ft 4in (8.94m); floatplane: approx 32ft 7in (9.93m)
Weight: 4,554lb (2,066kg)
Engine: 525hp Rolls-Royce Kestrel IB 12-cyl V-type
Max speed: 184mph (296kmh) at 5,000ft (1,525m)
Operational ceiling: 21,350ft (6,510m)
Range: 470 miles (756km)
Armament: 2×0.303in Vickers machine-guns
Max bomb load: 520lbs (236kg)

Sydney Camm's Hart 2-seat day bomber was one of the most technically and strategically significant bombers to emerge in the 1930s. With the Fairey Fox, it represented such a breakthrough in performance that it affected both British light bomber and fighter development. Air Ministry Specification 12/26 led Camm to use a powerful, low-drag, inline engine – the new Rolls-Royce Kestrel – to ensure the required marked performance improvement. The development of the all-metal structure, new methods of construction and the powerplant delayed the prototype's first flight until June 1928. A production contract was placed for 15 Harts; delivery began in January 1930. Home-based aircraft were replaced by Hinds from late 1935, but those in India served until 1939; it also served in Egypt and Palestine. British production, by Hawker, Armstrong Whitworth, Gloster and Vickers, totalled 984, approximately half being Hart Trainers. Hawker built 57 designated Hart (India); another tropical version was the Hart (Special). The standard 2-seat bomber was termed Hart Mk I or SEDB (Single-Engined Day Bomber). The total includes six Demons, eight Harts exported to Estonia in 1932 and four Pegasus radial-engined Harts sold to Sweden in 1934. A further 24 Pegasus Harts were built in Sweden in 1935-36, many fighting with the Finnish Air Force against the Soviet forces in 1939-40. Major variants were the Demon, Hind, Audax, Hardy, Hector and Osprey. The Hind day bomber (581 built by Hawker), an interim Hart replacement, introduced the Kestrel V and redesigned rear cockpit. The Demon (304 built by Hawker and Boulton-Paul) was a day fighter Hart, originally called the Hart Fighter. The Audax (618 built by Avro, Bristol, Gloster, Hawker and Westland), for the Army co-operation role, had long exhaust pipes and under-fuselage hook for message pick-ups. The general purpose Hardy (47 built by Gloster) had a similar hook, plus underwing racks for supply containers. The Hector (178 built by Westland), also an Army co-operation type, had an 805hp Napier Dagger H-type, changing the nose contours, and no sweep on the upper wings. The fighter-reconnaissance Osprey, the Hart's naval counterpart, with deck-landing or twin-float undercarriage, was first flown in mid-1930. Hawker built 136 for the FAA, one being sold to Spain, and four with Pegasus radials to Sweden; eight were built in Sweden. Hawker built four Hartbees, a developed Audax with Kestrel V, for the SAAF; 65 were built at Pretoria.

Hawker Fury

Hawker Fury Mk I flown by the Officer Commanding (OC), No 43 Sqdn, RAF, Tangmere, early 1932

Span: 30ft 0in (9.14m)
Length: 26ft 8in (8.13m)
Weight: 3,490lb (1,583kg)
Engine: 525hp Rolls-Royce Kestrel IIS 12-cyl V-type
Max speed: 207mph (333kmh) at 14,000ft (4,265m)
Operational ceiling: 28,000ft (8,535m)
Range: 305 miles (491km)
Armament: 2×0.303in Vickers machine-guns
Max bomb load: light bombs below wings

In the Hawker Fury, the epitome of British biplane fighter design in the 1930s, the lines of the famous Hurricane monoplane could already be seen. Historically, the Fury is also linked with Hawker's first production fighter, the Woodcock. A metal-framed fighter with an obvious Woodcock lineage, the Hawker Heron, was flown in 1925, but failed to secure an RAF production contract; and a similar fate met two subsequent designs, the Hornbill and the Hawfinch. Then, in August 1928, came the first flight of a new single-seat fighter, designed to the requirements of Air Ministry Specification F20/27. It had at first a bulky, uncowled Jupiter VII radial engine. In subsequent tests with a Mercury VI radial as its powerplant it reached a speed of 202mph (325kmh) at 10,000ft (3,050m). Upon this design Hawker based the Hornet, a most handsome biplane with a Rolls-Royce V engine, which was a centre of attraction at the 1929 Olympia Aero Show in London. The Air Ministry bought the Hornet prototype and, after trials at Martlesham Heath, Specification 13/30 was 'written round' this aircraft; 21 were ordered for the RAF and named Fury. Eventually, 118 Fury Mk Is were built for the RAF, entering service with No 43 Sqdn at Tangmere in May 1931 and serving later with Nos 1 and 25 Sqdns. Progressive improvements were made to G-ABSE, a Hawker-owned trials aircraft that became known as the Intermediate Fury, from which the High Speed Fury was developed, first flown on 3 May 1933. These two aircraft led to one of the standard Fury Mk Is (K1935) being brought up to the then current standard of G-ABSE with streamlined main-wheel fairings and a Rolls-Royce Kestrel VI engine. This underwent AAEE trials, after which Specification 6/35 was issued to cover production of 23 similar aircraft as Fury Mk IIs for the RAF. A further 75 were later ordered from General Aircraft Ltd, although not all of the latter were completed and some of them were diverted to the SAAF. Fury Mk IIs entered service in December 1936 with No 25 Sqdn, and later served with Nos 41, 43, 73 and 87. They had been phased out of these units by January 1939, but 16 Fury Mk Is and 48 Fury Mk IIs were serving with various Flying Training Schools upon the outbreak of World War 2. Several export Furies were built. Variants of the Mk I included 16 for Persia with Hornet radial engines, three for Spain, and six with Hispano-Suiza 12 NB engines for Yugoslavia, plus one Panther IIIA-engined aircraft evaluated by the Norwegian government. Counterparts of the Fury Mk II included six for Persia, three for Portugal and 10 for Yugoslavia.

PZL P-7a of No 111 Sqdn, 1st Air Regiment, Polish Air Force, Warsaw, *circa* autumn 1933

Span: 33ft 9½in (10.30m)
Length: 23ft 6in (7.16m)
Weight: 3,047lb (1,382kg)
Engine: 485hp Skoda-built Bristol Jupiter VIIF 9-cyl radial
Max speed: 200mph (322kmh) at 16,405ft (5,000m)
Operational ceiling: 32,810ft (10,000m)
Range: 435 miles (700km)
Armament: 2×0.303in Vickers, machine-guns

The P-7, one of the leading fighter aircraft of the inter-war years, was designed by Ing Zygmunt Pulawski. It stemmed from his P-1 fighter, one of the first designs undertaken by the Panstwowe Zaklady Lotnicze (National Aviation Establishments) upon their foundation in Warsaw in 1928. The P-1 had a high gull wing, an all-metal structure with metal covering, oleo-pneumatic main landing gear and a 600hp Hispano-Suiza V engine. The first P-1 prototype flew on 25 September 1929 and the second in 1930. From the P-1 were developed the radial-engined P-6 and P-7. The former had a Bristol Jupiter VI, the first of two prototypes flying in August 1930. The two P-7 prototypes, with 'high-altitude' Jupiter VIIFs, began flight testing in October 1930 and early spring 1931. Retaining basically the P-1's wings, tail and landing gear, both had redesigned oval-section fuselages. Ultimately the P-7 was selected. The production version followed the P-7/II second prototype in having a narrow Townend ring around the Skoda-built Jupiter instead of the P-7/I's individual cylinder fairings. Following an initial small order for production P-7as, contracts brought production to 150. Refinements included a narrow rear fuselage with cockpit head-rest fairing, and an improved Townend ring developed by PZL. Deliveries began in 1933, first to the famous *Eskadra Kosciuszkowska*. By the end of 1933, all first-line fighter squadrons of the Polish Air Force's 1st, 2nd, 3rd and 4th Air Regiments were equipped with the P-7a, making the PAF the first air force whose entire fighter force consisted of all-metal monoplanes. The P-7a remained in Polish service, alongside other and later designs, until the outbreak of World War 2, when there were still 30 operational with three fighter squadrons, and over 70 more on the strength of training establishments. Pulawski's excellent design formed the basis for the development, after his death in March 1931, of the more powerful P-11, by his successor, Ing Wsiewolod Jakimiuk. The first prototype, P-11/I, flew in September 1931. The initial production series (30 P-11a) had Skoda-built Mercury IVS2s. Delivery to PAF units began in 1934, and in 1935 50 P-11cs with 595hp Gnome-Rhône K9s went to Romania. The P-11c, in production in 1935, with improved view, became the standard Polish version (175 built), equipping 12 PAF squadrons at the outbreak of World War 2, P-11as having been withdrawn. The P-11 was versatile, and its handling made it popular with its pilots, but it was outclassed by its *Luftwaffe* opponents.

Caproni Ca 101

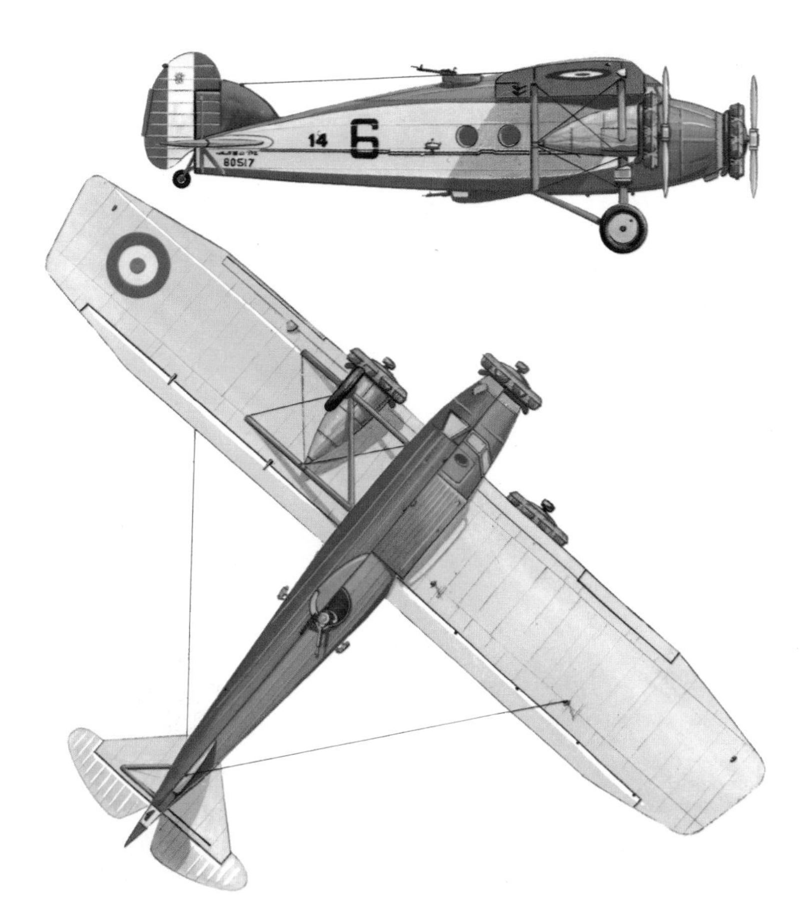

Caproni Ca 101D2 of the *Regia Aeronautica*, Italian East Africa, *circa* 1933

Span: 64ft 6¾in (19.68m)
Length: 45ft 3¼in (13.80m)
Weight: 10,968lb (4,975kg)
Engines: 3×240hp Alfa Romeo D2 9-cyl radials
Max speed: 103mph (165kmh) at 3,280ft (1,000m)
Operational ceiling: 20,000ft (6,100m)
Range: 621 miles (1,000km)
Armament: 2 or 3×7.7mm machine-guns
Max bomb load: 1,102lb (500kg) internally and externally

The Ca 101, itself a derivative of Caproni's earlier Ca 97, became the design basis for several subsequent trimotor designs, chief among which were the Ca 111 and the Ca 133. Like the somewhat later Ca 309, it also was employed primarily as a general-purpose 'colonial' aircraft for service in Italian overseas territories. It first entered service with the *Regia Aeronautica* a year or so before Italy's invasion of Ethiopia in 1936, when it was allocated a more belligerent role with the *Stormi da Bombardamento* as an attack aircraft, in addition to performing as a supply transport or ambulance. The Ca 101 was produced for civilian as well as military use, and in the former capacity was fitted with a wide variety of alternative powerplants during its career. Most of the military examples, however, were powered by the Alfa Romeo D2 engine, usually uncowled. The Ca 101 was a sturdily-built aircraft and its thick, broad wings in particular combined great strength and lifting ability. It had largely disappeared from *Regia Aeronautica* service by the end of the Ethiopian campaign, in favour of later developments of the design. Small numbers were built of the Ca 102, which was essentially a twin-engined counterpart of the Ca 101. One Ca 102 was completed with four engines mounted in tandem pairs, and was flown by the *62° Sperimentale Bombardieri Pesanti* (Experimental Heavy Bomber Squadron). The Ca 111 retained the essentials of the Ca 101 airframe, but despite its great size and weight was powered by a single engine only – the 950hp Isotta-Fraschini Asso 750RC radial. It appeared both in landplane form and as the Ca 111 *bis Idro* twin-float seaplane, the former serving alongside the Ca 101 in the Ethiopian campaign. It remained in service during World War 2, one example still surviving at the time of the Italian surrender in 1943. Ethiopia also saw the debut of the Ca 133, a slightly-enlarged development powered by 3 450hp Piaggio PVIIC14 radial engines in NACA-type cowlings. When Italy entered World War 2, 14 *Squadriglie da Bombardamento Terrestre* in East Africa were flying Capronis of this type, and others were used in the Italian invasion of Albania or as transports or ambulances in North Africa.

Junkers Ju 52/3m

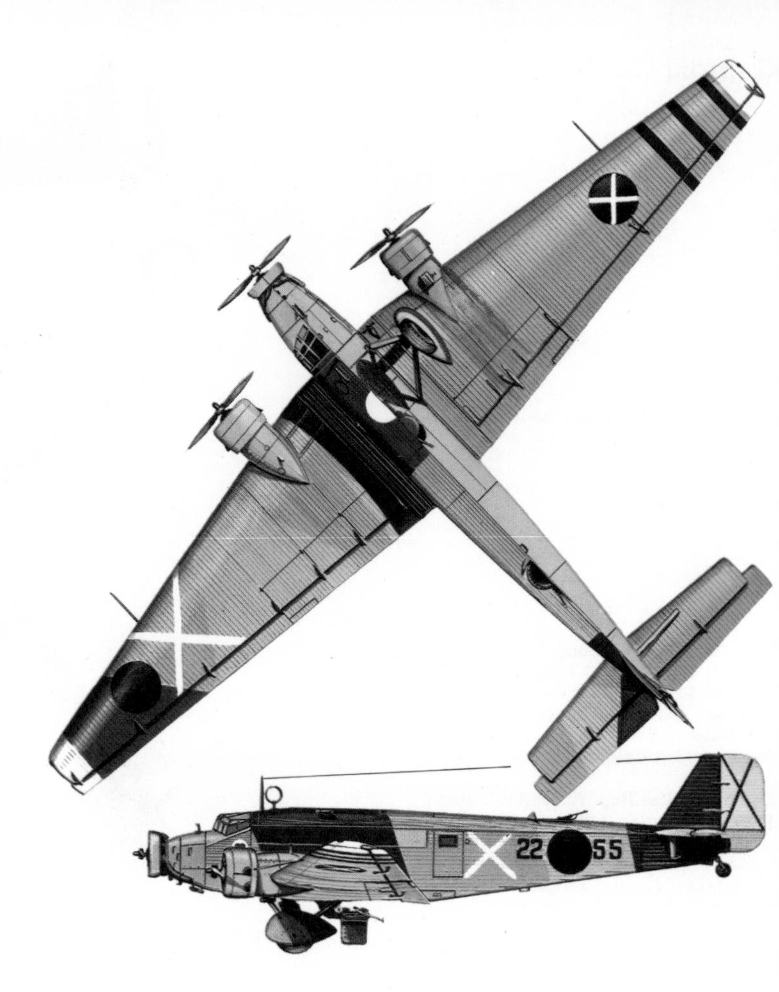

Junkers Ju 52/3m g3e of *Grupo 2-G-22,* Spanish Nationalist Air Force, *circa* 1936

Span: 95ft 11½in (29.25m)
Length: 62ft 0in (18.90m)
Weight: 20,944lb (9,500kg)
Engines: 3×600hp BMW 132A-3 9-cyl radials (licence-built Pratt & Whitney Hornets)
Max speed: 180mph (290kmh) at sea level
Operational ceiling: 20,670ft (6,900m)
Range: 795 miles (1,280km)
Armament: 2×7.9mm MG15 machine-guns
Max bomb load: 2,205lb (1,000kg), internally

The Junkers Ju 52/3m originated in 1928 with the Ju 52, which was powered by a 800hp Junkers L-88 and flew on 13 October 1930. Some six months later a development powered by 3 575hp BMW-built Pratt & Whitney Hornet (BMW 132) radials flew, designated Ju 52/3m; the single-engined version was retrospectively designated Ju 52/1m. The trimotor entered production in 1934, and quickly became a popular commercial transport with Deutsche Lufthansa and foreign operators, notably in Europe and South America. Adopted by the embryo *Luftwaffe* as interim equipment, 450 were built as Ju 52/3m g3e 4-seat bombers. They first saw operational service in late summer 1936 in the Spanish Civil War, when 20 were among the first German aircraft sent to Spain as bombers and troop transports. By late 1937 it had been superseded by the He 111 and Do 17, and largely transferred to troop transport duties, more appropriate to its capabilities. It retained its dorsal 7.9mm MG15 machine-gun, but the ventral gun and its retractable 'dustbin' mounting were often omitted. Manufacture in Germany continued until mid-1944, producing 2,193 by the end of 1939 and 2,650 thereafter. Production, including civil aircraft and French/Hungarian/Spanish licence manufacture totalled 4,851. Suffixes from g3e to g14e, distinguished military variants, signifying either the specific role, installation of later BMW 132 variants or structural alterations. Early production models had 600hp BMW 132As; the Ju 53/3m g5e introduced the higher-powered BMW 132T. Principal production model was the Ju 52/3m g7e, with enlarged cabin doors, autopilot and other improvements, fitted either as an 18-seat troop transport or a 12-stretcher ambulance. Popularly known as 'Iron Annie', nearly 600 were engaged in the invasion of Norway in April 1940, and of France and the Low Countries; in both campaigns well over a quarter of those involved were lost. Nevertheless, they were prominent in subsequent campaigns, notably in Greece, Libya and Crete. Duties included supply transport and glider tug; some had electromagnetic de-gaussing rings for clearing minefields. Most models had interchangeable wheel, ski or float landing gear. Many pre-war commercial Ju 52/3ms were impressed for military service. Postwar, several hundred were allocated to many foreign airlines, a few remaining in service in the 1970s. Postwar, Amiot in France built 100 as the AAC1, and CASA 100 equivalent to the Ju 52/3m g7e, for the *Ejército del Aire.* Known as the T2B in Spanish service, the CASA 352-L accommodated 18 fully-equipped troops or an equivalent freight load. The CASA 207 Azor replaced it from 1960 as Spain's standard transport, but several served as hack transports and for paratroop training for some years.

Fiat CR 32

Fiat CR32 of the *Aviación de Tercio,* Spanish Nationalist Forces, 1937

Span: 31ft 2in (9.50m)
Length: 24ft 5¼in (7.45m)
Weight: 4,112lb (1,865kg)
Engine: 600hp Fiat A30 RA 12-cyl V-type
Max speed: 233mph (375kmh) at 9,845ft (3,000m)
Operational ceiling: 29,530ft (9,000m)
Range: 466 miles (750km)
Armament: 2×0.303in Vickers (or 2× 12.7mm Breda-SAFAT) machine-guns
Max bomb load: 220lb (100kg)

The basic features of the Rosatelli-designed CR20 were retained in his next major design, the CR30, which was built in limited quantity for home-based units of the *Regia Aeronautica* and for the *2° Stormo* based in Libya. Small quantities were also built for Austria, China and Paraguay. The CR30 was powered by a 600hp Fiat A30 engine, had a 2-gun armament and a top speed of 217mph (350kmh). Compared with the CR20, it had a more cleanly-cowled engine, propeller spinner, streamlined wheel fairings and round-tipped wing and horizontal tail surfaces. The CR30 was followed in 1933 by the first flight of the further-improved CR32, about 350 of which were eventually built with 600hp Fiat A30 RA engines and an armament of 2 Vickers or Breda machine-guns in the top of the fuselage. The CR32 was a strong, agile little fighter, much liked by its pilots, and maintained the tradition set by its forebears by becoming the mount of the *Pattuglie Acrobatiche* aerobatic teams. In August 1936, CR32 fighters were sent to Spain to form the *Aviacion del Tercio,* which came to be recognised as one of the leading fighter formations on the side of Franco's Nationalist forces. During the Civil War, Italian and Spanish formations used 380 CR32s, known to the latter as the *Chirri.* The overall total of CR32 variants reached 1,212, the initial version being followed by the CR32*bis* which had the A30*bis* engine, 2 additional 7.7mm guns mounted in the wings and could carry one 220lb (100kg) or 2 110lb (50kg) bombs. Over 300 of this model were built, being supplanted on the production lines in 1936 by the CR32*ter,* of which about 100 were built. This model was basically similar to the CR32*bis* but introduced improvements in internal equipment and the landing gear. The last and most numerous version was the CR32*quater.* Both the CR32*ter* and *quater* reverted to the original armament of 2 guns. In performance and firepower the CR32 was outclassed as a fighter by the outbreak of World War 2, although with the later CR42 it still constituted some two-thirds of the available fighter strength of the *Regia Aeronautica.* But it was a highly versatile aeroplane, and continued to be used quite extensively during the early war years, chiefly in Greece and East Africa, for such duties as night fighting and close support. Foreign purchasers of the CR32*bis, ter* and *quater* included Austria, Bolivia, China, Hungary, Paraguay and Venezuela; in Spain, Hispano built 100 of the CR32*quater* as the HS-132L, later converting 40 to 2-seat fighter trainers.

Boeing P-26

Boeing P-26A of the 94th Pursuit Sqdn, 1st Pursuit Group, USAAC, *circa* 1935

Span: 27ft 11½in (8.52m)
Length: 23ft 10in (7.26m)
Weight: 2,955lb (1,340kg)
Engine: 600hp Pratt & Whitney R-1340-27 Wasp 9-cyl radial
Max speed: 234mph (377kmh) at 6,000ft (1,830m)
Operational ceiling: 27,400ft (8,350m)
Range: 635 miles (1,022km)
Armament: 2×0.30in (or 1×0.30in and 1×0.50in) Browning machine-guns
Max bomb load: 200lb (90kg)

Known affectionately as the 'Peashooter', the Boeing P-26 single-seat fighter was one of the more distinctive monoplane designs of the 1930s, combining several forward-looking features with others that, in retrospect, seem outmoded by comparison. It retained a well streamlined but nonetheless drag-creating fixed landing gear, yet at the time of its design (1931) Boeing was installing retractable gear in its B-9 bomber. At the same time, its low-wing monoplane layout and all-metal construction were truly modern features, even though the wings were somewhat extensively braced. It was, in fact, the first all-metal monoplane production American fighter. The P-26, Boeing's last production fighter, originated with three private-venture prototypes, known as the Boeing Model 248, designed and equipped with minimal assistance from the USAAC. Two of these aircraft were in due course evaluated in 1932 by the Army which in January 1933 ordered 111 examples of an improved version under the designation P-26A. As originally built, ailerons were the P-26As only wing control surfaces, but all aircraft in service later had flaps fitted. The first P-26As began to be delivered in December 1933, and they soon became standard pursuit squadron equipment in Hawaii and the Panama Canal area. A subsequent contract was placed for a further 25 fighters, two of these (designated P-26B and delivered in June 1935) having fuel injection engines. The remainder were delivered from February 1936 initially as P-26Cs, having the same powerplant as the A model and minor control system changes, but later many were converted to B standard. The P-26 served with USAAC pursuit and attack squadrons at home and overseas, but by 1940 had been replaced by Curtiss P-36s and Seversky P-35s and relegated to training. The P-26 had a high landing speed – alleviated by flaps – and poor soft-field landing characteristics, and was mainly used at bases with hard earth airfields in the south-western mainland and Hawaii. Many P-26As based overseas were sold to the Philippines and Panama; Guatamala purchased two, and others later from Panama. Boeing built only 12 of an export version, the Model 281, which differed only in detail from the P-26A (Model 226A), also having flaps added. Between 1934 and 1936 11 were exported to the Chinese Air Force which used them successfully against Japanese raiders. One was exported to Spain.

Handley Page Heyford

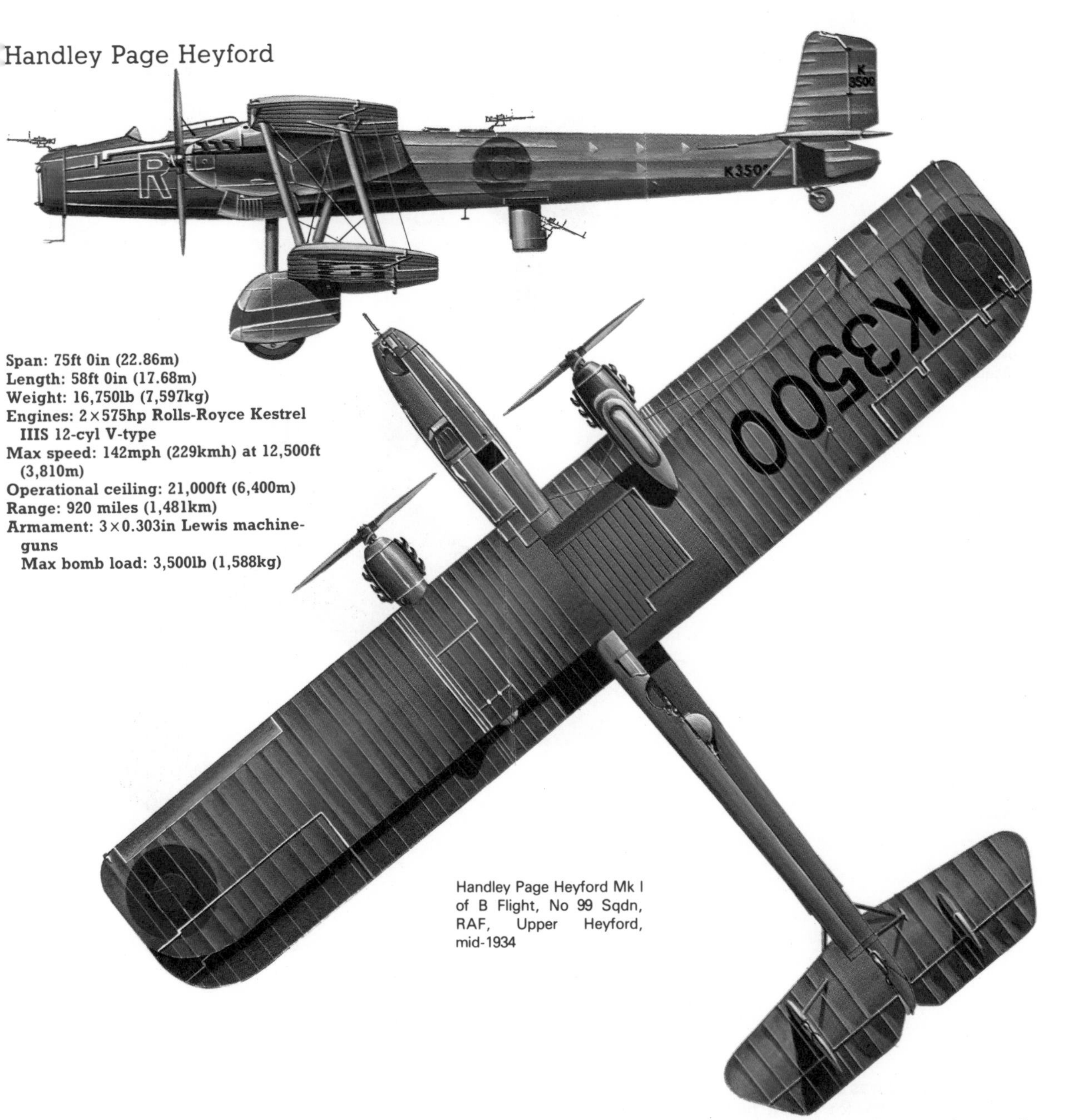

Span: 75ft 0in (22.86m)
Length: 58ft 0in (17.68m)
Weight: 16,750lb (7,597kg)
Engines: 2×575hp Rolls-Royce Kestrel IIIS 12-cyl V-type
Max speed: 142mph (229kmh) at 12,500ft (3,810m)
Operational ceiling: 21,000ft (6,400m)
Range: 920 miles (1,481km)
Armament: 3×0.303in Lewis machine-guns
Max bomb load: 3,500lb (1,588kg)

Handley Page Heyford Mk I of B Flight, No 99 Sqdn, RAF, Upper Heyford, mid-1934

No 99 Sqdn, at Upper Heyford, was appropriately the first RAF unit to be equipped with the Heyford 4-seat night bomber, which it began to receive in December 1933. The bomber had been designed to Specification B19/27, issued 6 years earlier, for a new type to replace the Hinaidi and Vickers Virginia. Two designs were accepted, the other being the Fairey Hendon monoplane, which was built in smaller numbers and entered service in late 1936. The HP 38 Heyford prototype (J9130) was designed under G R Volkert of Handley Page, and made its first flight at the company's airfield at Radlett, Hertfordshire, in June 1930. The Heyford's configuration was, for a biplane, extremely unorthodox. The upper wing was shoulder-mounted on the fuselage and bore the installation of the 2 Rolls-Royce Kestrel II engines. Heavy interplane struts then supported the lower mainplane at some considerable distance beneath the fuselage, and below this wing were the very substantial main landing wheels, protected by large fairings attached to the wing's leading-edge. This layout gave the Heyford a height on the ground of 20ft 6in (6.25m), allowing fitters and ground crew head-high clearance to walk beneath the front fuselage and permitting servicing to be carried out very quickly. The fields of view and fire enjoyed by the gunners and other crew members were exceptionally good, though care was obviously more than usually necessary when landing the aircraft. The first production model was the HP 50 Heyford Mk I, with open cockpit and Kestrel IIIS engines, and the similar Mk IA with a motor-driven instead of wind-driven generator; 38 Mk I/IAs were built to Specification B23/32, one Mk I (K3503) being completed with an enclosed crew cabin. The open cockpit was retained on the 16 Kestrel VI-powered Mk IIs which followed in 1934. The Mk II's engines were derated, but in the 70 Mk IIIs which completed production they developed their full 640hp each. The Heyford served with 11 first-line RAF bomber squadrons in the UK. From 1937, with the advent of monoplane bombers, it was relegated to training, serving until early World War 2.

Polikarpov I-15

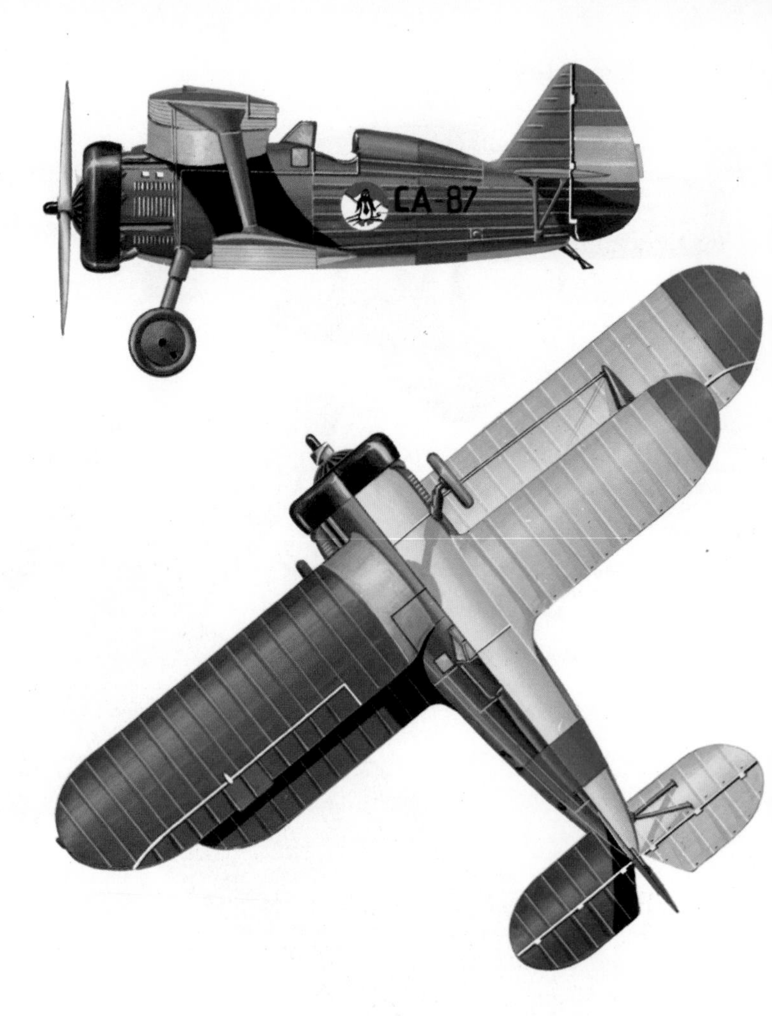

Polikarpov I-15 of the *4ª Chato Escuadrilla,* Spanish Republican Forces, 1937

Span: 31ft 11¾in (9.75m)
Length: 20ft 0½in (6.00m)
Weight: 4,189lb (1,900kg)
Engine: 715hp M-25 (licence-built Wright Cyclone) 9-cyl radial
Max speed: 229mph (368kmh) at 9,845ft (3,000m)
Operational ceiling: 32,150ft (9.800m)
Range: 310 miles (500km)
Armament: 4×7.62mm Nadashkevich PV-1 machine-guns

The Soviet I-15 fighter was a progressive development of the earlier I-5, design work for which began at the Aviatrust (TsKB) in Moscow early in 1933. The most noticeable innovations were the gull-pattern upper wings, with much-simplified strut-bracing, the single cantilever main landing gear legs, and the adoption of a more bulky radial engine, a Soviet version of the American Wright Cyclone. The prototype, designated TsKB-3, first flew in October 1933; rapid and successful completion of state trials enabled series production to begin in 1934, initial deliveries to service units starting late that year. Polikarpov's design team had also begun developing an improved version known as the I-15*bis* or I-152. It had a conventional straight upper wing carried above the fuselage, in deference to pilots' complaints that the gull-wing restricted the view forward. The wings were also increased in span and area, faster-firing guns were installed, and fuel capacity was increased. In 1936 the I-15*bis,* with an M-25V engine, began to enter service and the Soviet government sent the first I-15s to support the Republican forces in Spain. More than 500 I-15 and I-15*bis* fighters were dispatched to Spain, where, despite the abnormal numbers shot down, they remained popular. Substantial numbers of I-15*bis* operated on the USSR's eastern borders, fighting the Japanese on the frontiers with Manchuria and Mongolia. Here they met their match in the JAAF's Ki-27, but by this time, another development had been authorised, the I-15*ter* (or I-153). The prototype flew at the end of 1938 powered by an M-25V, but the production I-15*ter* had the more powerful 1,000hp Shvetsov M-62R engine. Developed by Artem Mikoyan, the I-15*ter* reintroduced the gull upper wing and had fully retractable main landing gear and was one of the fastest biplanes ever. The I-15*ter* was put very quickly into production, in a successful attempt to regain ascendancy over the Japanese. I-153 production continued until late 1940, the final model having a 1,100hp M-63, and served until 1944, despite its obsolescence and replacement by more modern fighters. Nearly 200 I-15*bis* and a substantial number of I-15*ter* were supplied to Nationalist China between 1937-40. Some captured aircraft, sold to Finland by Germany, fought against Soviet forces during 1942.

Heinkel He 51

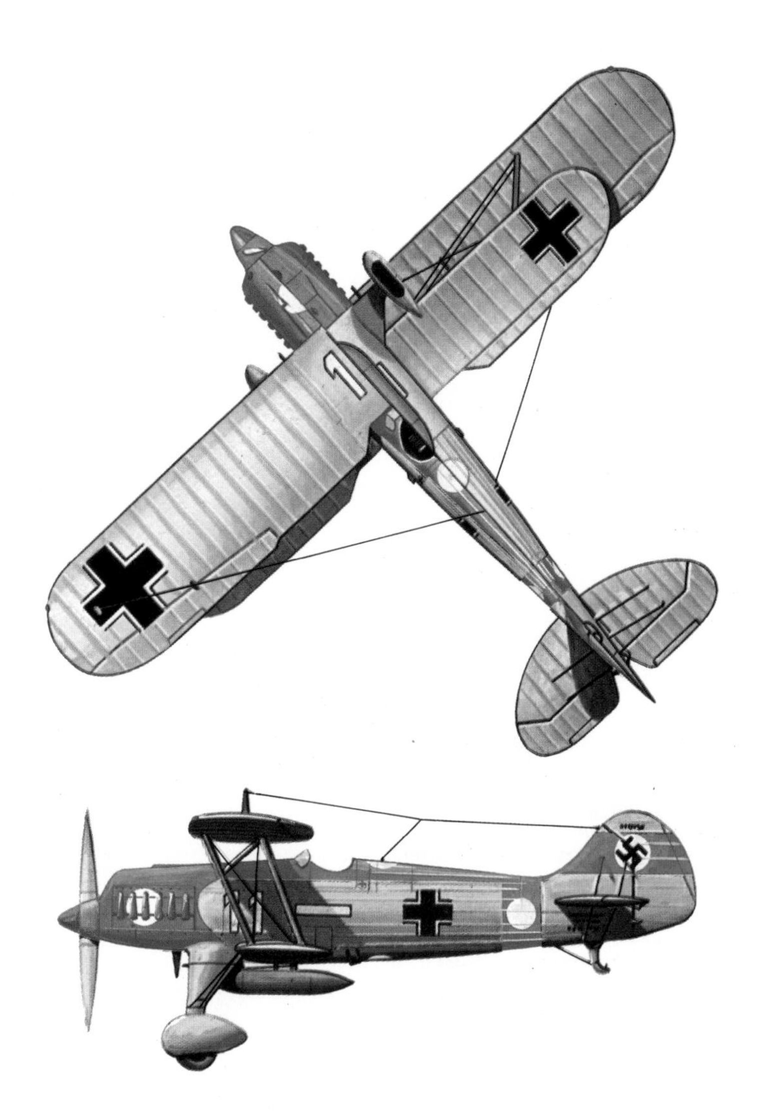

Heinkel He 51B-1 of *Staffel 6, II/JG Richthofen, Luftwaffe,* Döberitz, 1936

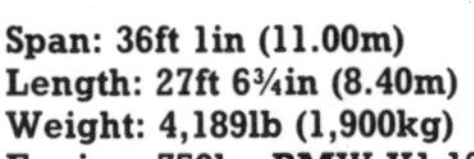

Span: **36ft 1in (11.00m)**
Length: **27ft 6¾in (8.40m)**
Weight: **4,189lb (1,900kg)**
Engine: **750hp BMW VI 12-cyl V-type**
Max speed: **205mph (330kmh) at sea level**
Operational ceiling: **25,260ft (7,700m)**
Range: **354 miles (570km)**
Armament: **2×7.9mm MG17 machine-guns**

The origins of the He 51 fighter are traceable through a series of small, streamlined fighters, beginning with the He 37 evolved in the late 1920s and continuing via the He 38, He 43 and He 49. It was the He 49a, designed by Siegfried and Walter Günter and first flown in November 1932, which effectively became the prototype for the He 51a, the fourth He 49a embodying sufficient modifications to justify the allocation of the later designation. This aircraft (D-ILGY) first flew in mid-1933; later that year construction began of 9 pre-series He 51A-0's for service trials, and delivery of these began in the summer of 1934. Delivery of the first production model, the He 51A-1, started in April 1935 to *JG 132,* and later to *JG 131* and *134.* Early in 1936 the next production model began to enter service. This was the He 51B-1, preceded by 12 pre-production B-0's and differing from its predecessor in having modified landing gear bracing and provision for an auxiliary ventral fuel tank. He 51B production included a number built as twin-float He 51B-2s for service at coastal fighter stations. At the end of July 1936, 6 He 51s, the first of many, were sent to Spain to support the Nationalist forces in the Civil War fighting under General Franco. Eventually, 135 He 51s served either with the Nationalist Air Force or Germany's *Legion Condor* in the Civil War. Flown by Spanish pilots, the first He 51s achieved some early success, but when, a few months later, Soviet Polikarpov I-15 biplane fighters began to appear in the Republican ranks, the German fighter found itself completely outclassed and suffered extensive losses. By the end of the war almost two-thirds of the He 51s fighting in Spain had been lost. Formations of He 51s continued to operate with some effect as ground attack aircraft, a role for which the He 51C was built from the outset: it was the He 51C which developed the 'cab-rank' system of ground-attack. Those supplied to the Franco forces were He 51C-1s, while those used by the *Luftwaffe* in Germany were He 51C-2s, featuring some equipment changes. But the disappointing results achieved as a fighter in Spain – reinforced by the He 51's poor showing in comparative trials at home with the Arado Ar 68 – led to its withdrawal from first-line *Luftwaffe* fighter units to close-support or training roles in 1938, after a comparatively short career. A total of 700 He 51 production aircraft was built. This included 75 A-1s by Heinkel, 225 (including 75 A-1s) by Arado, 200 by Erla and 200 (at least half of which were He 51Cs) by Fieseler.

Fairey Swordfish

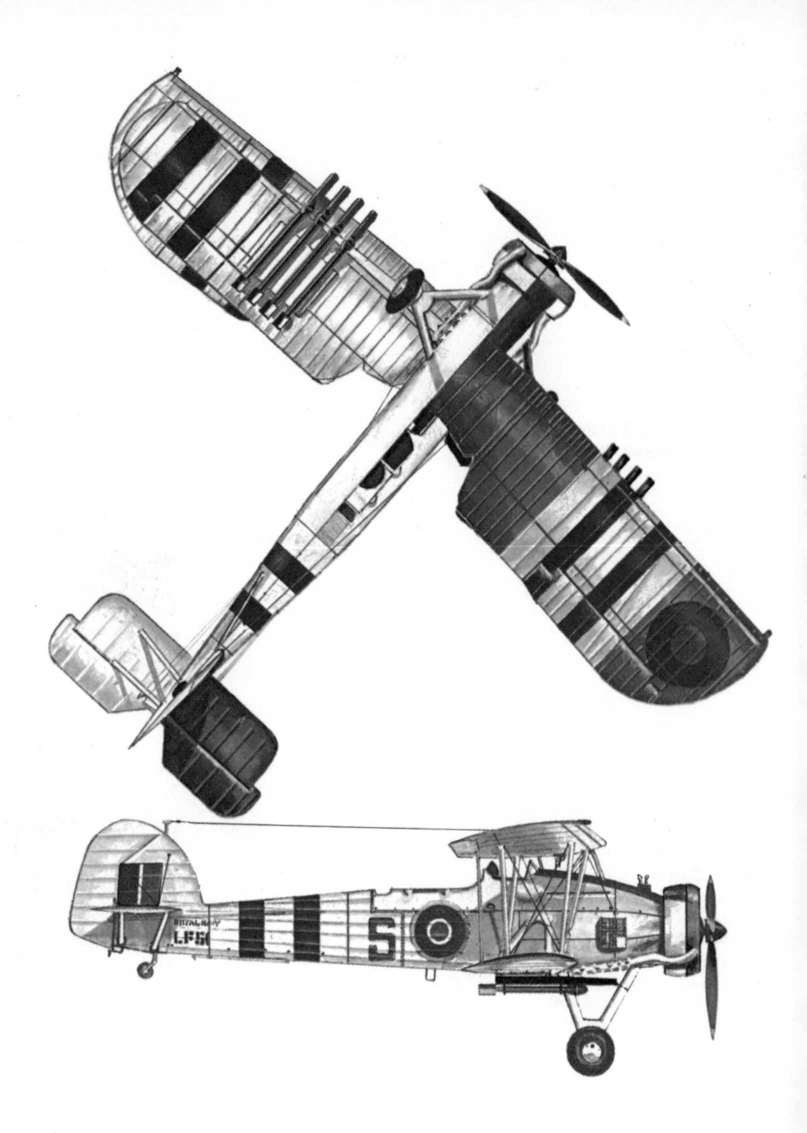

Fairey Swordfish Mk I of No 814 Sqdn, FAA, HMS *Ark Royal,* late 1938

Span: 45ft 6in (13.87m)
Length: 36ft 4in (11.07m)
Weight: 7,720lb (3,502kg)
Engine: 690hp Bristol Pegasus IIIM3 9-cyl radial
Max speed: 154mph (248kmh) at 7,000ft 2,134in)
Operational ceiling: 19,250ft (5.865m)
Range (with torpedo): 546 miles (879km)
Armament: 1×0.303in Vickers machine-gun; 1×0.303in Vickers K or Lewis machine-gun
Max bomb load: 1×18in torpedo, or 1,500lb (680kg) mine, or bombs

The Fairey Aviation Co had established a tradition of building sea-going aircraft when, in 1932, it designed the TSR I torpedo/spotter/reconnaissance biplane as a private venture. This prototype flew on 21 March 1933, but was lost in a crash in September. The design had already been proved satisfactory, and to Air Ministry Specification S15/33 Fairey built a modified prototype designated TSR II, the true Swordfish prototype; it had a lengthened fuselage, enlarged tail and revised landing gear. It flew on 17 April 1934, and an initial production batch of 86 Swordfish Mk I was ordered in 1935 to Specification S38/34, powered by the Pegasus IIIM3. Delivery to the FAA began in February 1936, first recipient being No 825 Sqdn on HMS *Glorious* in July. The Swordfish also undertook considerable training work. As an alternative to a 1,610lb torpedo carried beneath the fuselage, the Mk I could carry a 1,500lb mine in the same position or an equivalent weight of bombs under the fuselage and lower wings. After completing 692, including prototypes, Fairey ceased production early in 1940 to concentrate on the Albacore, but 1,699 later versions were built by Blackburn. At the outbreak of World War 2, 13 FAA squadrons were equipped with Swordfish, a figure later almost doubled. At first they were employed largely on convoy or fleet escort duties but they were soon employed upon minelaying. The first major torpedo attack was made in April 1940 by Swordfish from HMS *Furious* during the Norwegian campaign. Their most notable achievement was destroying three battleships, two destroyers, a cruiser and other warships of the Italian Fleet at Taranto on 10/11 November 1940, for the loss of two Swordfish. The Mk II appeared in 1943, with metal-covered lower wings, enabling it to carry RPs. Later Mk IIs had 820hp Pegasus XXXs which also powered the Mk III, identifiable by an ASV radar housing beneath the front fuselage. Swordfish from all three marks were converted to Mk IV, with enclosed cockpits, for the RCAF. A substantial number of Mk Is were converted as twin-float seaplanes for service aboard catapult-equipped warships, a configuration test-flown on 10 November 1934. The Swordfish was in action until less than four hours before the German surrender, but after VE-day was swiftly retired, the last squadron disbanding on 21 May 1945. Universally known as the 'Stringbag', its lengthy and successful career, outlasting the Albacore, its intended successor, was a tribute to its flying qualities, robustness and adaptability.

Consolidated PBY Catalina

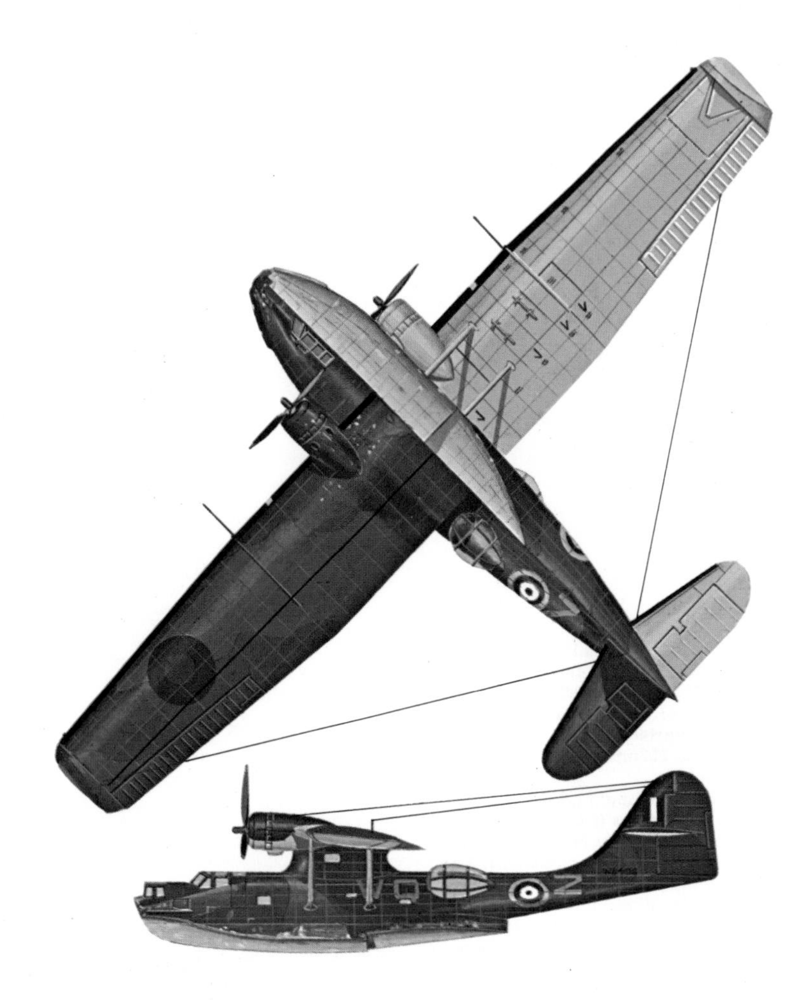

Consolidated Catalina Mk I (PBY-5) of No 209 Sqdn, RAF, which spotted the *Bismarck* on 26 May 1941

Span: 104ft 0in (31.70m)
Length: 63ft 10½in (19.47m)
Weight, normal take-off: 27,080lb (12,283kg)
Engines: 2×1,200hp Pratt & Whitney R-1830-S1C3-G Twin Wasp 14-cyl radials
Max speed: 190mph (306kmh) at 10,500ft (3,200m)
Operational ceiling: 24,000ft (7,315m)
Range: 4,000 miles (6,437km)
Armament: 6×0.303in Vickers machine-guns
Max bomb load: 2,000lb (907kg) externally

The Consolidated XP3Y-1, or Model 28, was one of two prototypes (the other being the Douglas XP3D-1) commissioned by the USN in October 1933 for comparative evaluation as patrol flying-boats. Isaac M Laddon's design for the Consolidated Aircraft Corpn showed extremely clean lines, particularly in its near-cantilever wing, which was mounted on a pylon above the fuselage and had stabilising floats which retracted to form the wing tips. The XP3Y-1 first flew on 21 March 1935, powered by two 825hp R-1830-54 Wasps; later that year (after a change to the Patrol Bomber designation XPBY-1) 60 production PBY-1s were ordered. These could carry up to 2,000lb of bombs, and had four 0.30in machine-guns. The XPBY-1 was delivered to VP-11F in October 1936, followed shortly by the first production aircraft. In 1937-38, 50 PBY-2s followed, and in 1938, three PBY-3s and a manufacturing licence were sold to the USSR. The Soviet version, designated GST, was powered by M-62s. Orders for the USN continued with 66 PBY-3s (R-1830-66s) and 33 PBY-4s (1,050hp R-1830-72s), the latter introducing the prominent waist blisters that characterised most subsequent versions. The RAF received one Model 28-5 for evaluation in July 1939, resulting in an order for 50 similar to the USN's PBY-5, which had 1,200hp R-1830-82s or -92s and a redesigned rudder. The RAF name Catalina was subsequently adopted for USN PBYs. During 1940 the RAF doubled its order; other Catalinas were ordered by Australia (18), Canada (50), France (30), and the Netherlands East Indies (36). Of the USN's original order for 200 PBY-5s the final 33 were completed as PBY-5A amphibians, and an additional 134 were ordered to PBY-5A standard; 12 later became RAF Catalina Mk IIIs and 12 more were included in the Dutch contract. A total of 753 PBY-5s was built and 794 PBY-5As, 56 of the latter being completed as OA-10s for the USAAF. Lend-Lease supplies to Britain included 225 PBY-5Bs (Mk IA) and 97 Mk IVAs with ASV radar. Production continued with the tall-finned Naval Aircraft Factory PBN-1 Nomad (156, most going to the USSR) and the similar PBY-6A amphibian (235, including 75 USAAF OA-10Bs and 48 for the USSR). Canadian Vickers-built amphibians went to the USAAF (230 OA-10As) and RCAF (149, named Canso). Boeing Canada production included 240 PB2B-1s (mostly as RAF Mk IVBs), 17 RCAF Catalinas, 50 tall-finned PB2B-2s (RAF Mk VI) and 55 RCAF Cansos. US/Canadian PBY production totalled 3,290; several hundred GSTs were built in the USSR. The Catalina's performance enabled it to maintain its viability with many military and naval air arms, as well as a commercial transport, for many years after the war, particularly in South America. Those operating Catalinas for maritime reconnaissance up to the mid-1960s included Argentina, Brazil, Chile, Ecuador and Mexico, while Nationalist China, Dominica, Indonesia and Peru still used the type for SAR, and France and Israel retained a few for miscellaneous duties. By 1970, however, very few military or civil Catalinas remained in service.

Heinkel He 111 (CASA 2.111)

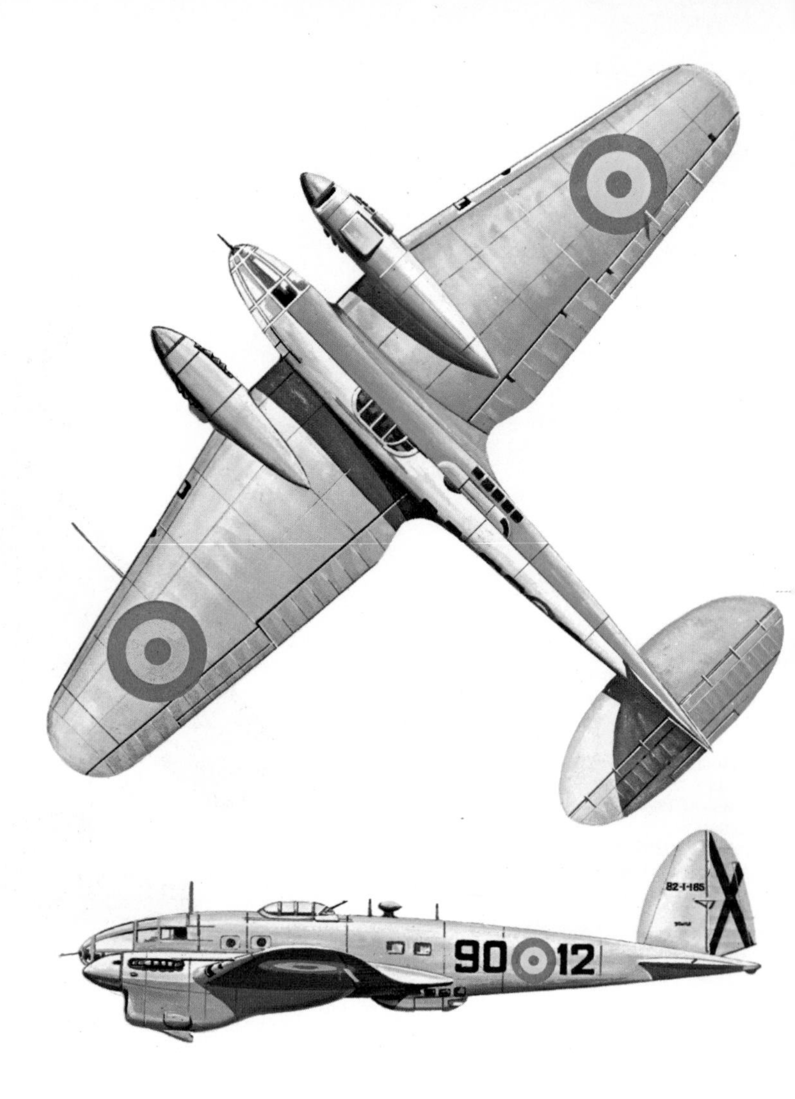

CASA 2.111D (B21) of the *Aviación Tactica,* Spanish Air Force, *circa* 1965

Span: 74ft 1¾in (22.6m)
Length: 53ft 9¾in (16.4m)
Weight: 30,865lb (14,000kg)
Engines: 2×1,350hp Junkers Jumo 211F-2 12-cyl inverted V-type
Max speed: 252mph (405kmh) at 19,685ft (6,000m)
Operational ceiling: 21,980ft (6,700m)
Range: 1,280 miles (2,060km)
Armament: 1×20mm MG FF cannon; 1×13mm MG 131 machine-gun; 2-4× 7.9mm MG 81 machine-guns; 1-2× 7.9mm MG 15 machine-guns
Max bomb load: 4,410lb (2,000kg) internally; or 1,764lb (800kg) internally plus 2,205lb (1,000kg) externally; or 5,510lb (2,500kg) externally (NB: data apply to He 111H-16)

The He 111's service career lasted over 30 years, an outstanding tribute to the design, developed by the Günter brothers early in 1934. The first of four prototypes flew on 24 February 1935; two were completed ostensibly as civilian transports. In summer 1935 a pre-series batch of He 111A-0s appeared but their BMW engines provided insuffient power. The first major type was the He 111B, with DB600 engines, which was one of the Condor Legion's most successful types in Spain. To preserve DB600s for fighters, the He 111D did not enter large-scale production, the next major versions being the Jumo-powered E and F. The latter had the revised, straight-tapered wings evolved for the proposed commercial G. E and F models served in Spain, and after the Civil War nearly 60 surviving He 111s joined the Spanish Air Force. Production up to F reached nearly 1,000 by the outbreak of World War 2. A new model then appeared, the He 111P, whose extensively glazed, restyled nose, with offset ball turret, became an He 111 characteristic. It was built in comparatively small numbers, to conserve Daimler-Benz engines, but its Jumo-powered counterpart, the He 111H, became the most widely used series, well over 5,000 being built before production ended in 1944. Reflecting Battle of Britain experience, subsequent types appeared with progressive increases in armament. Although most extensively used in the intended role of medium bomber, He 111Hs carried a variety of operational loads. The H-6 was particularly effective as a torpedo bomber, while other H sub-types became carriers for the Hs 293 glider bomb and the V1 flying bomb. The H-8 had balloon-cable cutters. The He 111Z *Zwilling* (Twin) glider tug was two H-6 airframes linked by a new wing centre-section with a fifth Jumo. The H-23 was an 8-seat paratroop transport. Total German He 111 production exceeded 7,000. CASA built 200 H-16s, designated C.2111, under licence in Spain, the first flying in 1945; the first 130 had Jumos but, when supplies ran out, the remainder were completed with Rolls-Royce Merlins; many Jumo-powered bombers were converted to Merlins. The 2.111D was a Merlin-powered reconnaissance bomber; 15 bombers were converted to 2.111E 9-seat military transports; and the 2.111F was a dual-control trainer. *Ejército del Aire* designations were: B.2H for Jumo-engined bombers; B.21 for Merlin bombers; and T.8 for transports. Latterly relegated to aircrew training and miscellaneous duties, a number survived in Spanish service until the late 1960s.

Dornier Do 17

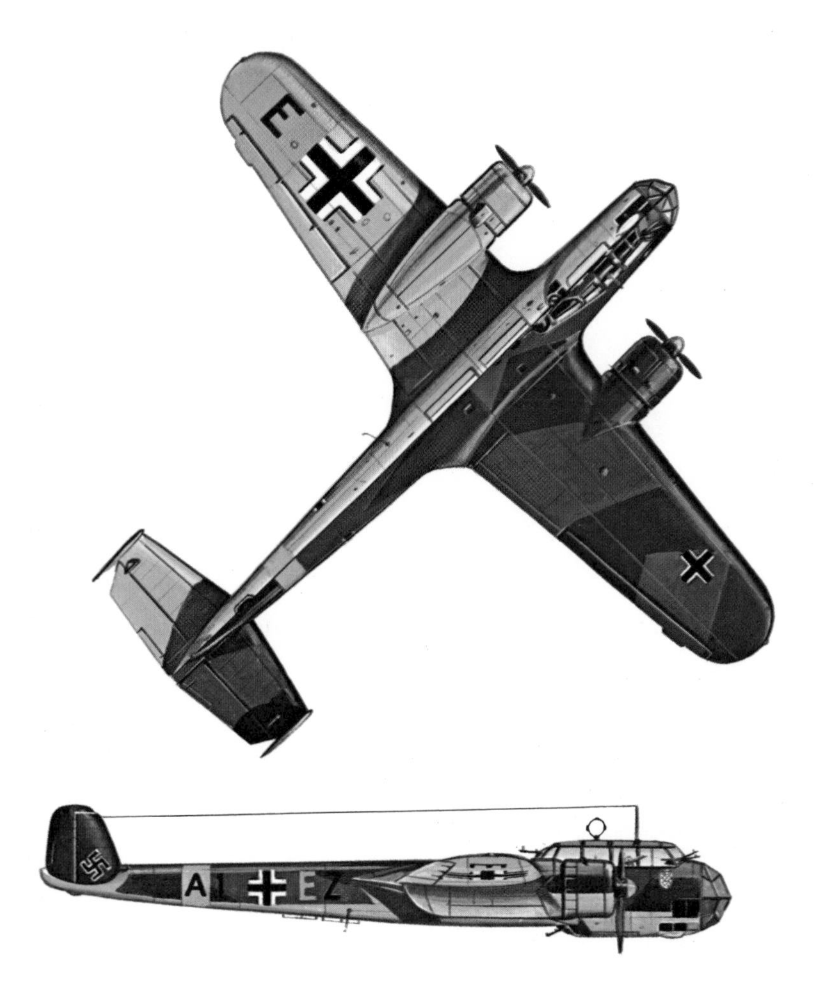

Dornier Do 17Z-2 of 15/*KG*.53 *Kroaten, Luftwaffe,* Eastern Front, September 1942

Span: 59ft 0⅝in (18.00m)
Length: 51ft 9⅝in (15.79m)
Weight, normal take-off: 18,872lb (8,560kg)
Engines: 2×1,000hp Bramo Fafnir 323P 9-cyl radials
Max speed: 224mph (360kmh) at 13,125ft (4,000m)
Operational ceiling: 22,965ft (7,000m)
Range with 1,102lb (500kg) bomb load and auxiliary fuel: 721 miles (1,160km)
Armament: 6×7.9mm MG15 machine-guns
Max bomb load: 2,205lb (1,000kg) internally

Three prototype Do 17s, with single fins and rudders and 660hp BMW VI V-types, were built to meet a requirement by Deutsche Lufthansa for a high-speed mail and passenger aircraft, the first flying in autumn 1934. The extremely slim fuselage was unsuitable for a commercial transport. Later, the RLM evaluated the design as a medium bomber, ordering additional prototypes, the first, the Do 17V4, appearing with the twin fins that became standard. The fuselage, shorter than on the original trio, and wings, were of metal construction. The V4, V6 and V7 had BMW VIs: the V5 had 770hp Hispano-Suiza 12Ys. Production began late in 1936, first models being the Do 17E-1 bomber (750hp BMW VIs) and Do 17F-1 long-range reconnaissance bomber, both in service by mid-1937. Yugoslavia ordered 20 Do 17Ks, an export version with 986hp Gnome-Rhône 14N radials, which improved speed and range; this was later licence-built in Yugoslavia. In 1938 Do 17s were sent to the *Luftwaffe*'s Condor Legion in the Spanish Civil War. The Do 17M bomber (900hp Bramo Fafnir 323As) and the reconnaissance bomber Do 17P (865hp BMW 132N radials) entered production in late 1937, replacing the E and F respectively. The Do 17U pathfinder bomber (DB600As) entered service in 1938. The Do 17Z (Bramo 323As or 1,000hp Bramo 323Ps) featured a more angular, bulbous front fuselage; over 500 were built for bombing, reconnaissance and training. They were followed by 101 Do 215Bs; the B-5 was a night fighter intruder with a six-gun 'solid' nose. Development continued as the Do 217, the first of several prototypes flying in August 1938, resembling a scaled-up Do 215B. First major series was the Do 217E (BMW 801 radials) which entered production in 1941. The Do 217K (1,700hp BMW 801Ds) bomber series had a redesigned and even more bulbous nose. Inline engines – 1,750hp DB603As – appeared in the Do 217M. Final production model, a reconnaissance version, was the Do 217P (DB603As); only six were built. To overcome a severe shortage of specialised night fighters, the *Luftwaffe* in 1942 initiated large-scale conversion of Do 217 bombers, replacing the bulbous, glazed nose with a streamlined 'solid' fairing containing guns. First version converted was the Do 217E-2; 157 were redesignated Do 217J-1, an intruder with reduced bomb-carrying capacity, or J-2, a night fighter with no bomb bay, carrying Lichtenstein airborne interception radar. The Do 217N-1 and N-2 (DB603As) night fighters were similar to the J-2; 50 were converted from Ms in 1943. Total Do 217 production reached 1,730, all except 364 being bomber models.

Mitsubishi G3M (Nell)

Mitsubishi G3M3 Model 23 of the Genzan Naval Air Corps, JNAF, 1941

Span: 82ft 0¼in (25,00m)
Length: 53ft 11¾in (16.45m)
Weight: 17,857lb (8,100kg)
Engines: 2×1,300hp Mitsubishi Kinsei 53 14-cyl radials
Max speed: 258mph (416kmh) at 19,355ft (5,900m)
Operational ceiling: 33,725ft (10,280m)
Range: 2,090 miles (3,363km)
Armament: 1× 20mm cannon; 2×7.7mm machine-guns
Max bomb load: 1,764lb (800kg), internally

The JNAF competitions of 1934 were the first to produce significantly modern equipment comparable with the best western designs, in particular Mitsubishi's A5M fighter and G3M twin-engined medium bomber. Mitsubishi's association with Junkers in the early 1930s was reflected in the G3M, which resembled the Ju 86. The temporary designation Ka-15 was given to the project, which was based on the Ka-9, and designed by Professor Kiro Honjo. A cantilever mid-wing monoplane with a circular-section fuselage, flush-riveted skin and retractable main landing gear, the first Ka-15 prototype, powered by two 750hp Hiro Type 91 V-type engines, was completed in July 1935 and flew shortly afterwards. A further 20 prototype/pre-production Ka-15s were built, 17 with Mitsubishi Kinsei 2 or 3 radials of 680 or 790hp, which improved performance. The JNAF designation G3M1 was sub-sivided unofficially into G3M1a (Hiro engines), G3M1b (Kinsei engines) and G3M1c (those with glazed nose areas). Twelve G3M1s were used for service trials during which an initial production batch of G3M1s with Kinsei 3 engines was ordered; 34 were built before being supplanted by the G3M2, the principal version. Mitsubishi built two basic variants of the G3M2: the Model 21, with higher-powered Kinsei engines, increased fuel and modified dorsal turrets; and the Model 22, with revised armament installation, including eliminating the retractable ventral turret. Mitsubishi built 343 Model 21s between 1937-39, and 238 Model 22s between 1939-41. Nakajima also produced G3M2s, continuing after Mitsubishi ceased production in favour of the G4M. Nakajima built 412 G3Ms, including G3M3s with 1,300hp Kinsei 53 engines and further increased fuel capacity. The G3M entered JNAF service early in 1937, equipping two squadrons when the Sino-Japanese conflict began that July. They made world-wide news with transoceanic raids from Omura, Japan, and Taipei, Formosa, on the Chinese cities of Hangkow and Nanking; but poor fighter opposition misled the Japanese into thinking that bombers could operate safely without fighter escort. Additional fighters were summoned urgently from Japan, but in 1940 G3Ms themselves escorted the first two squadrons of A6M2 fighters to China! Upon the outbreak of the Pacific War in 1941 the JNAF had a first- and second-line strength of approximately 250 G3M2s; the type took part in the initial attacks on the Philippine Islands, and in sinking HMS *Prince of Wales* and *Repulse.* Before the war some G3M2s were converted to civil transports, and during the war a number of G3M1s and G3M2s became L3Y1 and L3Y2 military transports.

Gloster Gladiator

Gloster Gladiator Mk I of the Latvian Air Force, 1937

Span: 32ft 3in (9.83m)
Length: 27ft 5in (8.36m)
Weight: 4,592lb (2,083kg)
Engine: 830hp Bristol Mercury IX 9-cyl radial
Max speed: 253mph (407kmh)
Operational ceiling: 32,800ft (10,000m)
Range: 430 miles (692km)
Armament: 4×0.303in Browning machine-guns

When Air Ministry Specification F7/30 was issued for a four-gun day and night fighter with a speed of 250mph (402kmh), Gloster was developing the Gauntlet fighter. Seven British manufacturers submitted designs, but many foundered on the unsuitability of the Rolls-Royce Goshawk around which they were designed. Gloster pursued a private-venture developed version of the Gauntlet as a rival; based upon the 645hp Mercury VIS and designated SS37, it was flown in September 1934. The following spring it was acquired by the Air Ministry (RAF serial K5200) for evaluation. In June 1935, an initial contract for 23 was placed to Specification F14/35, powered by the 840hp Mercury IXS, with four Vickers guns (the prototype had had two Vickers and two Lewis) and an enclosed cockpit, as the Gladiator Mk I. Deliveries, initially with Lewis guns in the lower wings, were made in February/March 1937, the first aircraft being allocated to No 27 Sqdn, Tangmere. In September 1935 a second contract had been placed for 108 Gladiators; two years later eight home RAF fighter squadrons were fully operational with the type. Gladiators were the last biplane fighters to serve with the RAF, but monoplane fighters did not eclipse Gladiators as first-line fighters. During the Munich crisis in 1938, 13 of the 18 fully-operational UK fighter squadrons had Gladiators or Gauntlets. On 3 September 1939 the RAF had 210 Mk Is and 234 Mk IIs on charge, the latter figure including 38 Sea Gladiators. The Mk II differed from the Mk I in having an 840hp Mercury VIIIA, desert eqiupment and detail improvements; the Sea Gladiators were interim conversions from RAF machines, with catapult points and an arrester hook, and a ventral fairing for a collapsible dinghy. A further 60 Sea Gladiators were built for the FAA. Production amounted to 448. Gladiators were exported widely before World War 2, to Belgium (22), China (36), Eire (4), Greece (2), Latvia (26), Lithuania (14), Norway (12), Portugal (15), South Africa (11) and Sweden (55). Ex-RAF machines were supplied to Egypt (45), Finland (30), Greece (23), and Iraq (14), although several were later returned to the RAF. Gladiator production ended in 1940. By the Battle of Britain, only one UK fighter squadron was flying Gladiators. They had virtually disappeared from British service by mid-1941 but continued to serve in North Africa and on the Eastern Front, eventually being relegated to second-line duties, particularly meteorological.

Messerschmitt Bf 109

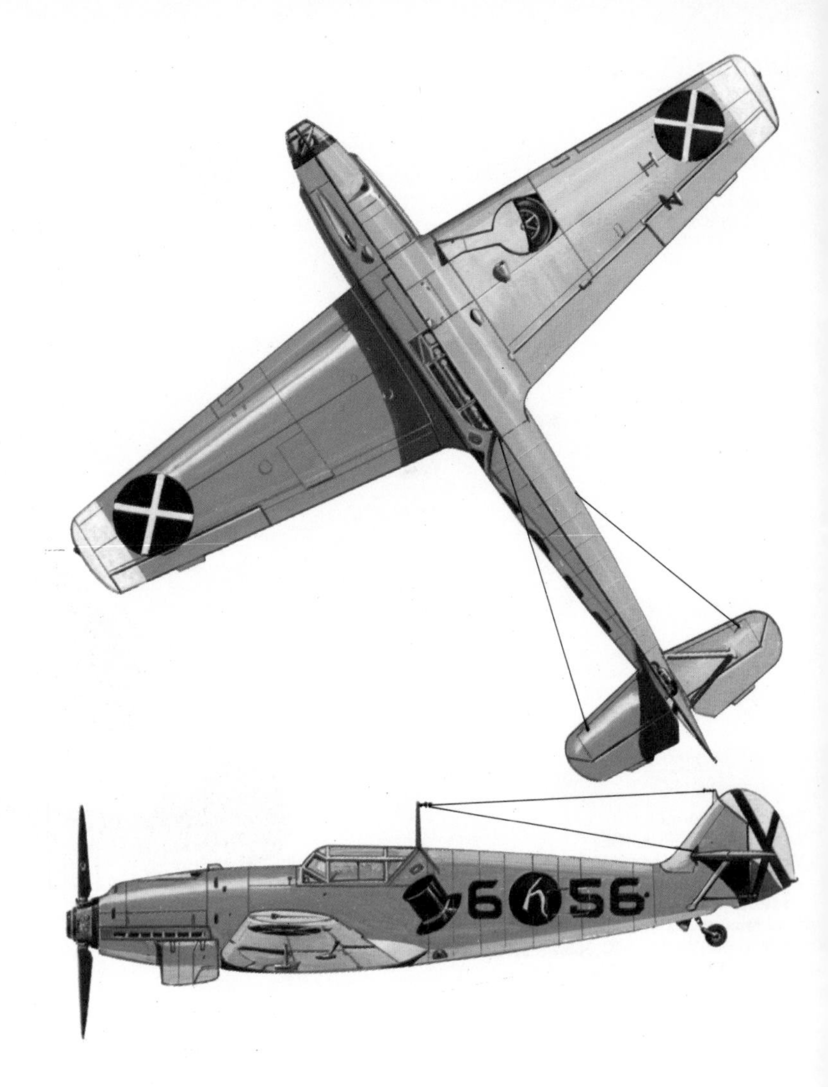

Messerschmitt Bf 109B-2 of *Staffel* 2, *J/88*, Condor Legion, Spain, 1937-38

Span: 30ft 3in (9.22m)
Length: 25ft 6in (7.77m)
Weight: 4,850lb (2,200kg)
Engine: 700hp Junkers Jumo 210G 12-cyl inverted V-type
Max speed: 292mph (470kmh) at 13,125ft (4,000m)
Operational ceiling: 26,575ft (8,100m)
Range: 429 miles (690km)
Armament: 3×7.9mm MG17 machine-guns
(NB: weight and performance data are for a B-1 with 650hp Jumo 210Da)

The Bf 109, the Bayerische Flugzeugwerke's response to a 1933 RLM specification for a fighter, was designed by Willy Messerschmitt's team around the 610hp Junkers Jumo 210A, the most powerful German aero-engine. This engine was not available for the first flight of the Bf 109V1 first prototype which flew in September 1935 with a 695hp Rolls-Royce Kestrel V, but was installed in the second prototype, which flew the following January. By then the RLM had ordered ten Bf 109s for evaluation. The Bf 109V3 was the prototype for the intended initial production model, the Bf 109A, but its two MG17 machine-guns were considered inadequate. Armament was increased to three on the actual first production version, the Bf 109B-1 (635hp Jumo 210D). By mid-1937 this was replacing the obsolete He 51 in service. In 1937 the Bf 109 attracted world attention at the International Flying Meeting at Zurich; then on 11 November 1937, the Bf 109V13, with a boosted DB601, set up a new world landplane speed record of 379.38mph (610.55kmh). In 1937 24 Bf 109B-2s joined the *Luftwaffe's* Condor Legion, fighting in the Spanish Civil War, followed by other Bs. Although successful as a combat aircraft, the B was comparatively lightly armed, and four or five guns were introduced in the Bf 109C-1, which appeared in Spain in 1938. In 1938, Bf 109 orders were such that five other companies began to contribute. Some B-2s were re-engined with DB600As, and redesignated Bf 109D-0, and followed in production by the similarly-powered D-1. At the outbreak of World War 2, 2,235 D-series were in *Luftwaffe* service, but were already being replaced by the Bf 109E which first appeared in 1938. This series was superior to virtually every fighter opposed to it in 1939 and early 1940. Bf 109E production mounted rapidly, and it was the principal version during the Battle of Britain. The series, extending to the E-9, included fighter, fighter-bomber, and reconnaissance models. The finest model was the F, a much cleaner design, with a 1,200hp DB 601N or 1,300hp DB601E, neater nose contours, cantilever tailplane and retractable tailwheel, and increased-span wings with round tips. By late summer 1942, the F series had been supplanted by the G, the last major production model. Intended to be an improved Bf 109F, its heavier DB605 and equipment reduced performance; yet production increased, and it was employed in Europe, North Africa and on the Russian front. The Bf 109K, built in small numbers, was a refined G. Licence production continued postwar in Czechoslovakia and Spain, producing an approximate total of 35,000.

Junkers Ju 87

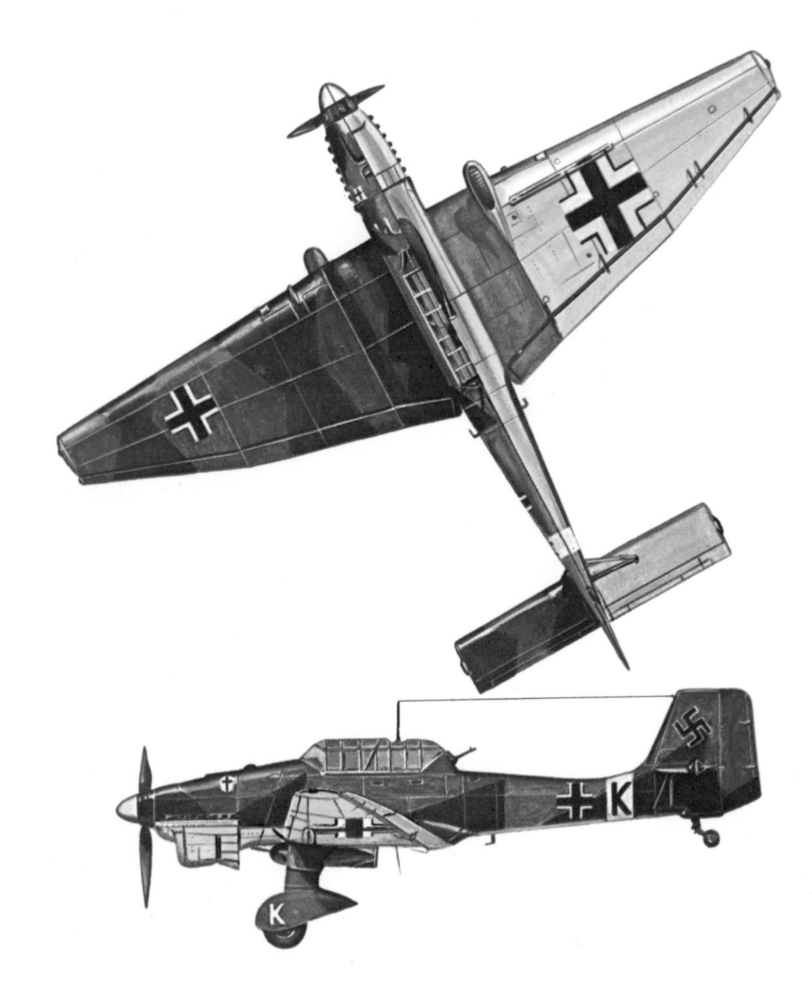

Junkers Ju 87B-1 of *Geschwader Stab/StG* 2 *Immelmann, Luftwaffe,* France, *circa* summer 1940

Span: 45ft 3¼in (13.80m)
Length: 36ft 5in (11.10m)
Weight: 9,370lb (4,250kg)
Engine: 900hp Junkers Jumo 211A-1 12-cyl inverted V-type
Max speed: 242mph (390kmh) at 13,410ft (4,400m)
Operational ceiling: 26,250ft (8,000m)
Range: 342 miles (550km)
Armament: 2×7.9mm MG17 machine-guns; 1×7.9mm MG15 machine-gun
Max bomb load: 1,102lb (500kg) externally

During the later 1930s and the early part of World War 2, the USA and Germany, in particular, were proponents of the dive bomber, typified in the Ju 87, design of which, by Dip-Ing Pohlmann, started in 1933. The first prototype flew early in 1935, powered by a Rolls-Royce Kestrel. It had rectangular twin fins and rudders, but the Ju 87V2, flown that autumn, had a single tail and a 610hp Junkers Jumo 210A, and was representative of production aircraft. A pre-series batch of Ju 87A-0s was started in 1936 and in spring 1937 delivery began of the initial production A-1, followed by the similar A-2. About 200 A series were built before, in autumn 1938, the Ju 87B appeared, with the Jumo 211 and enlarged tail, redesigned vertical tail and restyled landing gear fairing. A and B models served with the Condor Legion in Spain in 1938. By the outbreak of World War 2 the A had been relegated to training; 336 Bs were in front-line service. *Luftwaffe* fighter superiority eased the Ju 87's passage in 1939-40, but during the Battle of Britain losses were considerable. Nevertheless, B production continued into 1941, and it served in the Mediterranean and North Africa. Substantial numbers were supplied to Italy, Bulgaria, Hungary and Romania. Produced alongside the B was the long-range Ju 87R, used for anti-shipping and other missions from 1940. Before the war a few Ju 87C-0s were completed, a 'navalised' version for the abortive carrier *Graf Zeppelin;* they served with land-based units; production C-1s were completed as B-2s. The next major production variant was the Ju 87D, having a more powerful Jumo, increased fuel tankage and considerably refined airframe with increased armament and crew armour. Most D variants were produced for ground attack, carrying various weapon loads, ranging from a 1,800kg bomb to 12 7.9mm machine-guns in underwing pods. The D-5 introduced an extended wing of 49ft 2½in (15.00m) span. D's served in the Mediterranean, North Africa and on the Eastern Front, equipping Hungarian, Romanian and *Luftwaffe* units. The proposed Ju 87F and Ju 187 were abandoned in 1943. The anti-tank Ju 87G entered service in 1943 with two underwing 37mm BK37 cannon or bombs for ground or armour attack missions. A conversion of the D-5, it was successful on the Eastern Front until better Soviet fighters appeared. Operational ground attack trainers were produced, designated Ju 87H, by converting Ds. Production ended in September 1944 and over 5,700 Ju 87s were built.

Savoia-Marchetti SM79 Sparviero

Savoia-Marchetti SM79-II Sparviero of the 59° *Squadriglia,* 33° *Gruppo,* 11° *Stormo Bombardamento Tenestre, Regia Aeronautica,* Cyrenaica, November 1940

Span: 69ft 6⅝in (21.20m)
Length: 53ft 1¾in (16.20m)
Weight, normal take-off: 25,133lb (11,400kg)
Engines: 3×1,000hp Piaggio PXI RC40 14-cyl radials
Max speed: 295mph (475kmh) at 13,120ft (4,000m)
Operational ceiling: 27,890ft (8,500m)
Range: 1,243 miles (2,000km)
Armament: 3×12.7mm Breda-SAFAT machine-guns; 1×7.7mm Lewis machine-gun
Max bomb load: 2,756lb (1,250kg) internally

The SM79, designed by Alessandro Marchetti, originated as the prototype (I-MAGO) of an 8-seat commercial transport, making its first flight in October 1934. Several record flights, with various powerplants, were made in 1935-36, leading to the completion of the second prototype as a military bomber, powered by three 780hp Alfa Romeo 126RC34 engines. This entered production as the SM79-I Sparviero, (Hawk), but it was its dorsal hump which caused it to be nicknamed *il Gobbo* (the hunchback) when it entered service. The 8° and 111° *Stormi Bombardamento Veloce* (High Speed Bomber Groups) achieved some considerable success with their Sparvieri during the Spanish Civil War, and 45 SM79-Is were ordered in 1938 by Yugoslavia. In 1937 service trials were begun of the SM79-I equipped to carry one 450mm (17.7in) torpedo, and later two, beneath the fuselage. These trials indicated that, with more powerful engines, the Sparivero could easily carry two of these weapons externally, and in October 1939 production began of the SM79-II, to equip the *Squadriglie Aerosiluranti* (Torpedo Bomber Squadrons) of the *Regia Aeronautica.* Apart from one batch with 1,030hp Fiat A80 RC 41 engines, all SM79-IIs were powered by the Piaggio PXI radial. Production was sub-contracted to Macchi and Reggiane factories, and in June 1940 there were nearly 600 Sparvieri, of both models, in Italian service. The SM79 was active during World War 2 throughout the Mediterranean area and in North Africa and the Balkan states, its duties including torpedo attack, conventional bombing, reconnaissance, close-support and, eventually, transport and training. After the Italian surrender in 1943 about three dozen flew with the co-belligerent air force, while the pro-German *Aviazione della RSI* employed several SM79-IIIs. This was a cleaned-up version, without the ventral gondola, and had a forward-firing 20mm cannon. The SM79B was a twin-engined export model, with an extensively glazed nose, and was first flown in 1936. It was built for Brazil (three), Iraq (four) and Romania (48), each version with a different powerplant. The Romanian IAR factories also built the SM79B under licence with Junkers Jumo 211D inline engines. Overall Italian production of SM79 variants, including export models, reached 1,330 before output ceased in 1944.

Vickers Wellesley

Vickers Wellesley of the RAF Long Range Development Unit, flown by Sqdn Ldr R Kellett, one of three taking part in the world distance record flight, 5-7 November 1938, covering 7,157.7 miles (11,519 km). (Data are for standard Wellesley Mk I.)

Span: 74ft 7in (22.73m)
Length: 39ft 3in (11.96m)
Weight: 11,100lb (5,035kg)
Engine: 925hp Bristol Pegasus XX 9-cyl radial
Max speed: 264mph (425kmh) at 19,680ft (6,000m)
Operational ceiling: 33,000ft (10,060m)
Range: 2,590 miles (4,168km)
Armament: 1×0.303in Vickers machine-gun; 1×0.303in Vickers K machine-gun
Max bomb load: 2,000lb (907kg) (in underwing containers)

By any standards, the Wellesley was a remarkable aeroplane. Aircraft designers did not in the early 1930s normally think of long-range bombers in terms of elegant single-engined monoplanes with enclosed cockpits, cantilever wings and retractable landing gear. But two men did — Barnes Wallis and Rex Pierson, whose ideas stemmed from the geodetic construction used to built the airship R.100. They applied this to a single-engined biplane, designed to Specification M1/30, which flew in 1933; Wallis developed it further in the aircraft selected by the Air Ministry as Vickers' contender to Specification G4/31, for a new general-purpose and torpedo bomber. Simultaneously, however, Vickers proceeded with a private venture contender to G4/31 with a more fully-developed geodetic structure and long-span cantilever monoplane wings. This first flew on 19 June 1935, but the Air Ministry had decided to order 150 of the biplane bomber. However, the disparity in performance between the two aircraft led in September to a new order superseding the earlier contract, calling for 96 G4/31 monoplanes, to be named Wellesley. The prototype (serial K7556) underwent numerous modifications, including changing the Pegasus IIM3 engine for a Pegasus X, increasing the rudder area and adding streamlined underwing bomb containers. Geodetic construction was used for the thick-section high-aspect ratio wings, and the fuselage aft of the wing front spar. In April 1937, No 76 Sqdn at Finningley became the first RAF unit to equip with the Wellesley Mk I; other squadrons followed; Nos 35, 77, 148 and 207 in the UK and Nos 14, 45 and 223 in the Middle East. The UK Wellesley units re-equipped with twin-engined types in 1939, but in Africa the Wellesley served in the early part of the war. A total of 176 production Wellesleys was built. Of these, some later aircraft, often termed unofficially Mk II, had a continuous 'glass-house' canopy over front and rear cockpits. In 1938 five (L2637, '38, '39, '80 and '81) were converted for the RAF's Long Range Development Unit, powered by special Pegasus XXIIs, stripped of military equipment, and carrying 1,290 Imp galls (5,8641) of fuel instead of the normal maximum of 425 Imp galls (1,9321). In July, four of these aircraft flew non-stop from Cranwell to Ismailia in preparation for an attack on the world record for the greatest distance flown in a straight line. The attempt was made by L2638, '39 and '80 which took off from Ismaila on 5 November 1938 to reach Australia and landed at Darwin nearly 48 hours after leaving Egypt.

Hawker Hurricane

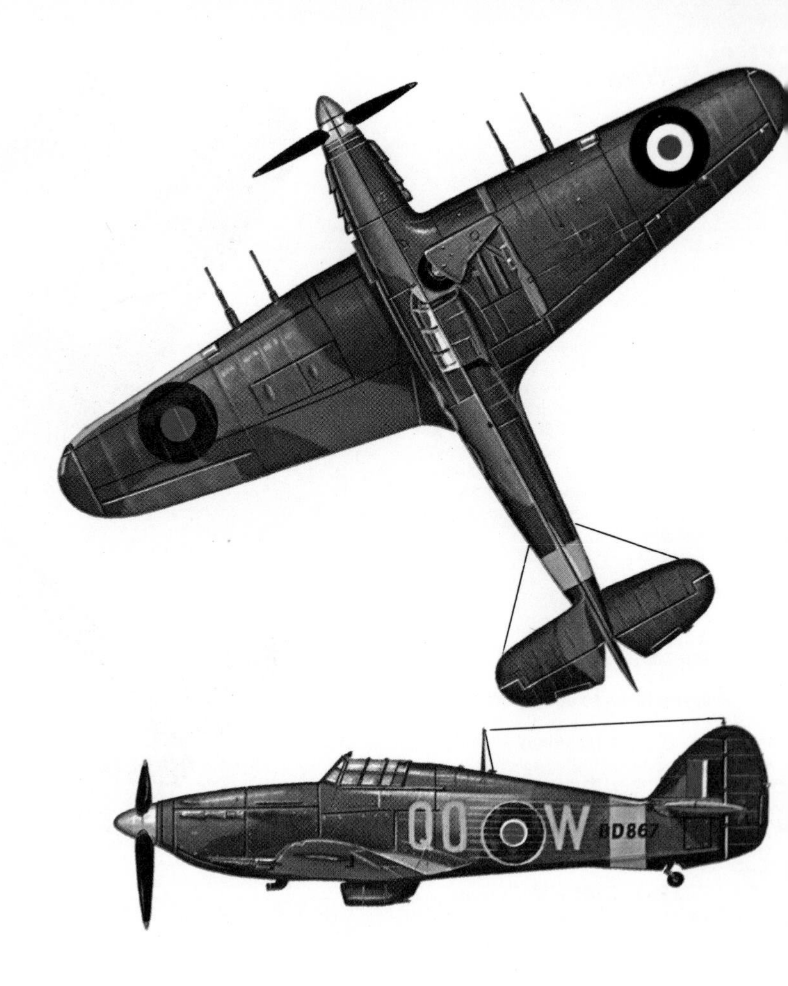

Hawker Hurricane Mk IIC of No 3 Sqdn, RAF, UK, 1941

Span: 40ft 0in (12.19m)
Length: 32ft 2¼in (9.81m)
Weight, normal take-off: 7,544lb (3,422kg)
Engine: 1,300hp Rolls-Royce Merlin XX 12-cyl V-type
Max speed: 329mph (529kmh) at 18,000ft (5,480m)
Operational ceiling: 35,600ft (10,850m)
Range, normal: 460 miles (740km)
Armament: 4×20mm Oerliken or Hispano cannon
Max bomb load: 1,000lb (454kg)

The first Hawker monoplane fighter, the Hurricane, is indissolubly associated with its designer, Sydney Camm. Its evolution began in 1933 as a monoplane development of the Fury biplane fighter, with a fixed landing gear and a Rolls-Royce Goshawk. Early in 1934, significant improvements included a fully retractable main undercarriage, eight wing-mounted 0.303in machine-guns and the use of the new Rolls-Royce PV-12 – the Merlin. To Specification F36/34, Hawker completed an unarmed prototype (K5083), powered by a 990hp Merlin C, which flew on 6 November 1935. In March 1936, without an official order, Hawker confidently began to prepare for production of 1,000, foresight emphasised during the Battle of Britain. Their confidence was justified: three months later the Air Ministry placed an initial order for 600 Mk Is. Delivery began to No 111 Sqdn, in October 1937, delayed slightly by the decision to use the Merlin II. On 3 September 1939 there were 315 on the fully operational strength of 14 RAF squadrons, plus 107 in reserve. Total orders stood at 3,500 of which nearly 500 had been completed. Mk I production by Hawker and Gloster factories in the UK and the Canadian Car and Foundry Co (160), totalled 3,954. Canadian-built machines were later redesignated Mk X. The Hurricane became operational with the Advanced Air Striking Force in France at the outset of World War 2, and outnumbered the Spitfire by about two to one in the Battle of Britain in August-October 1940. During the second half of 1940 the Mk I appeared in the Middle East, following Italy's entry into the war; these (and later Hurricanes in North Africa) had a Vokes sand filter beneath the nose. In 1942, Hurricanes appeared in Singapore, the Netherlands East Indies and Burma, operating also in the fighter-bomber role. On 11 June 1940, the prototype Mk II flew with a 1,185hp supercharged Merlin XX. Early production aircraft, with eight machine-guns, were designated Mk IIA and Mk IIB with 12 machine-guns; the Mk IIC had four 20mm cannon. The Hurricane was becoming outclassed as a fighter and in 1942 was increasingly used for ground-attack. Several Mk IICs were equipped with underwing RPs. The anti-tank Mk IID had two 40mm underwing cannon and two Brownings in the wings. The last British production model, the ground-attack Mk IV, had various weapon arrangements and a 1,620hp Merlin 24 or 27. The Sea Hurricane, with catapult spools and (except for the first 50 Mk IAs) an arrester hook, appeared in 1941 as an interim step to protect convoys, carried aboard CAM ships (Catapult Aircraft Merchantmen). Hurricane conversion resulted in Sea Hurricanes Mks IA, IB, IC and XIIA. The Mk IIC had a 1,460hp Merlin XX and full FAA radio and other equipment, but no catapult spools. About 800 Sea Hurricanes were built or converted. UK Hurricane production totalled 13,080 by Hawker, Gloster and Austin Motors; the Canadian Car and Foundry Co built a total of 1,451 Mks X, XI, XII and XIIA, with Packard-Merlins; overall production totalled 14,533. The Soviet Air Force was allocated 2,952 Hurricanes during early war years, although many were lost en route.

Fiat G50 Freccia

Fiat G50*bis* of the *151° Squadriglia, 20° Gruppo, 51° Stormo, Regia Aeronautica,* Libya, *circa* November 1941

Span: 36ft 1⅛in (11.00m)
Length: 27ft 2⅜in (8.29m)
Weight, normal take-off: 5,512lb (2,500kg)
Engine: 870hp Fiat A74 RC 38 14-cyl radial
Max speed: 302mph (486kmh) at 19,685ft (6,000m)
Operational ceiling: 35,270ft (10,750m)
Range, normal: 294 miles (473km)
Armament: 2×12.7mm Breda-SAFAT machine-guns

The G50, designed by Ing Giuseppe Gabrielli, was one of six designs (another was the Macchi C200) submitted to the Italian Air Ministry in 1936 for an all-metal, single-seat fighter monoplane with retractable landing gear. The first prototype (MM334) flew on 26 February 1937, powered by an 840hp Fiat A74 RC38 radial engine. An initial order for 45 G50s was given to the Fiat subsidiary CMASA, at Marina di Pisa, and delivery began in January 1938. Immediately, 12 of this batch were sent to Spain to join the Italian *Aviazione Legionaria* fighting with the Republican forces. Their participation was too short for a conclusive evaluation of the G50's combat worth, but production was maintained as an insurance against possible difficulties in Macchi C200 production. A further 200 G50s were ordered, and a total of 118 G50s were on *Regia Aeronautica* strength when Italy entered World War 2 in June 1940. Late in 1939 the Finnish government ordered 35, most of which were delivered by the spring of 1940, and these gave excellent service until May 1944 before being withdrawn. With the *Regia Aeronautica* the Freccia served in interceptor, ground attack, convoy and bomber escort roles, and was employed in Belgium, Greece and the Balkan theatre as well as in the Mediterranean and North Africa during the first half of the war. Fiat-CMASA built 245 of the original G50 before this version was supplanted in production by the improved G50*bis,* whose prototype was flown on 9 September 1940. This model had a refined fuselage design, modified canopy, a shorter and broader rudder, and increased fuel tankage and protective armour. Fiat's CMASA and Turin factories completed 421 of this version. CMASA also designed and built 108 G50B tandem 2-seat unarmed trainer models. Nine G50*bis* were delivered to the Croatian Air Force. Experimental variants tested included the DB601-engined G50V of 1941, discarded in favour of the G55 design, and the G50*bis*-A, an enlarged 2-seat fighter-bomber with increased armament and bomb load, flown in 1942 but abandoned after the Italian Armistice. The proposed G50*ter* was rendered abortive when the A76 engine intended for it was abandoned before the aircraft's first flight in July 1941.

Nakajima Ki-27 (Nate)

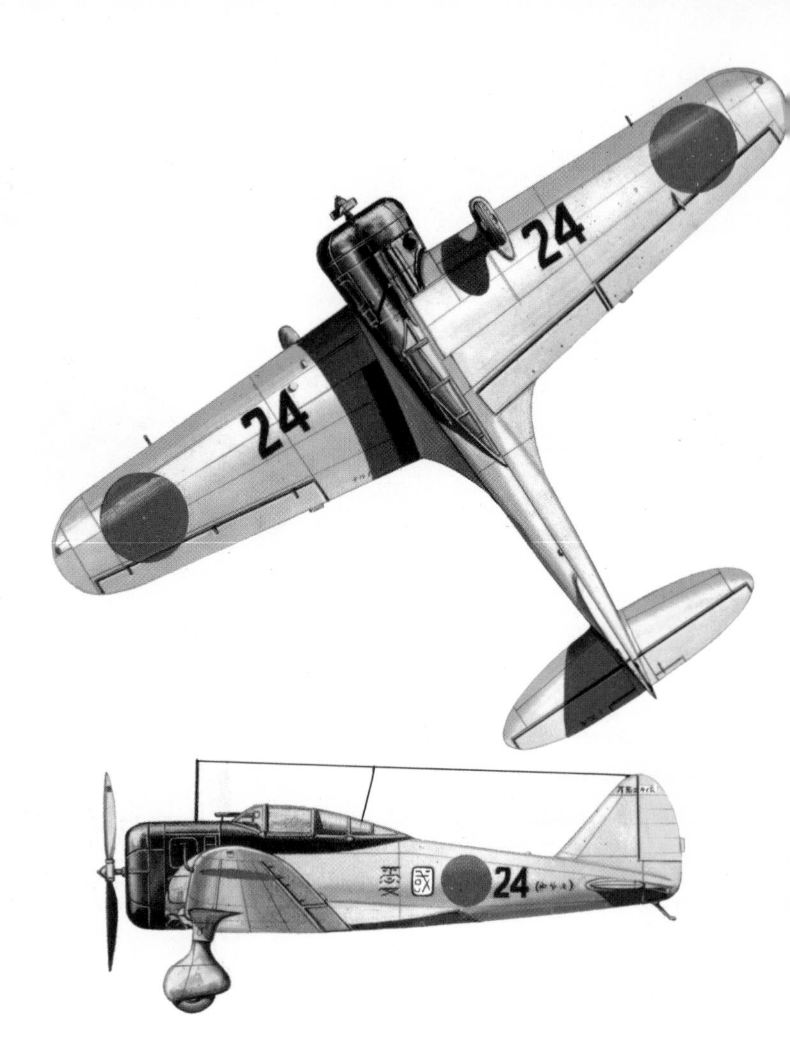

Nakajima Ki-27a donated to the JAAF (unit unknown), *circa* 1938-41

Span: 37ft 1in (11.30m)
Length: 24ft 8½in (7.53m)
Weight: 3,638lb (1,650kg)
Engine: 710hp Nakajima Ha-1b 9-cyl radial
Max speed: 286mph (460kmh) at 11,485ft (3,500m)
Operational ceiling: 9,845ft (3,000m)
Range: 388miles (625km)
Armament: 2×7.7mm Type 89 machine-guns
Max bomb load: 220lbs (100kg)

The Ki-27 was one of three designs developed to meet a 1935 JAAF requirement for a single-seat fighter. The other contenders were the Mitsubishi Ki-18 and the Kawasaki Ki-28. The Ki-28 had a faster speed and rate of climb, but the Nakajima design was selected for its lightness and higher degree of manoeuvrability. Three Ki-27 prototypes were completed, the first making its maiden flight on 15 October 1936, all powered by a 650hp Nakajima Ha-1a radial engine, and were identical except for progressively greater wing area. The JAAF chose the largest-area wing and ordered the manufacture of 10 pre-series Ki-27s and, in mid-1937, full-scale series production of the Ki-27a, or Type 97 Model A, powered by a 710hp Ha-1b. The Ki-27a thus became the JAAFs first low-wing monoplane, and its first fighter to have an enclosed cockpit. Deliveries began in February/March 1938, going first to JAAF fighter squadrons in Manchuria, which had previously flown Kawasaki Ki-10 biplanes. They made their combat debut against Soviet Air Force fighters in the Siberian border disputes of 1938-39. Others were flown by the Manchurian Air Force. The arrival of the Ki-27a in this area brought a swift, though temporary, change in JAAF fortunes, for it was highly successful against the Soviet I-5 fighters which opposed it. The I-15 and the monoplane I-16 proved more worthy opponents, but nevertheless the Ki-27as accounted for a substantial proportion of the 1,252 Soviet aircraft which the JAAF claimed during the campaigns; the JAAF lost about 100 Ki-27as. Detail improvements, including a modified canopy, were introduced on the Ki-27b (Type 97 Model B), which replaced the Ki-27a in production in 1939. A total of 2,007 Ki-27as and 1,379 Ki-27bs were built, the parent company producing 2,079 of this total; the remainder were manufactured by the Tachikawa Aircraft Co and the Manchurian Aircraft Co. Although virtually obsolete by Pearl Harbor, the Ki-27 was in widespread JAAF service; it was allotted the Allied code name Nate. Encountered in Burma, China and Malaya during the first six months of the Pacific war, it was gradually withdrawn for conversion to a 2-seat advanced training role with a 450hp engine under the designation Ki-79. Three examples were completed of the Ki-27-Kai in 1940, but this improved model was abandoned when the new Ki-43 Hayabusa became available.

Curtiss P-36/Hawk 75

Curtiss P-36C of the 27th Pursuit Sqdn, USAAC, in colours adopted for 1936 'War Games'

Span: 37ft 4in (11.38m)
Length: 28ft 6in (8.69m)
Weight: 5,800lb (2,631kg)
Engine: 1,200hp Pratt & Whitney R-1830-17 Twin Wasp 14-cyl radial
Max speed: 311mph (500kmh) at 10,000ft (3,050m)
Operational ceiling: 33,700ft (10,270m)
Range: 820 miles (1,320km)
Armament: 1×0.50in and 3×0.30in machine-guns

The Curtiss Hawk monoplane fighter was one of four contenders for a 1935 USAAC design competition. Its design, as the Curtiss Model 75, had begun in 1934, the prototype aircraft (X17Y) first flying in May 1935, the month planned for holding the USAAC competition, but none of the other contenders was ready and it was deferred. Improvements in all candidates were recommended after initial evaluation, the Curtiss design re-emerging as the Model 75B with modified cockpit and tail surfaces and a Wright Cyclone in place of the R-1760 fitted originally. Although the Seversky SEV-1XP fighter was the competition winner, three variants of the Curtiss 75B, with Twin Wasp engines, were ordered for trials, designated Y1P-36. Evaluation of these after their delivery in February 1937 led to a substantial production order for 177 P-36As and 31 P-36Cs, the latter having two additional wing guns and different variant of Twin Wasp engine. Deliveries to the USAAC began in April 1938. P-36s remained in US operational service into World War 2; a few, indeed, were engaged against the Japanese force that attacked Pearl Harbor on 7 December 1941. A novel feature of the P-36 was its method of landing gear retraction whereby the main wheel legs pivoted to enable the wheels to lie flat in the wings. Curtiss produced a simplified model for export in 1937, having an 875hp Wright Cyclone engine and a faired, non-retractable landing gear. Presentation of the prototype to the Chinese Air Force was followed by an order for 112 similar aircraft, designated Hawk 75M and delivered in 1938; 25 generally similar 75Ns were built for Thailand, and 30 75Os for the Argentine Air Force; the FMA factory in Argentina built a further 100 of this model. Exports were also made of retractable-undercarriage models. The first orders came from France, which from May 1938 placed order totalling 730 for Hawk 75A series fighters with revised armament, some with Twin Wasp and some with Cyclone engines. Delivery was still incomplete when France surrendered in June 1940, and the remaining aircraft were diverted to serve the RAF, which named them Mohawk. Other 75A series Hawks were ordered by Iran (10 with Cyclone engines), Norway (24 Twin Wasp Hawks and 36 with Cyclones) and the Netherlands (35 Cyclone Hawks), and HAL in India built 5 under licence. The German invasion interrupted deliveries of the Norwegian Hawks, and 24 of the Dutch aircraft were diverted to the Netherlands East Indies, where they fought against the Japanese in the Pacific war.

Nakajima B5N (Kate)

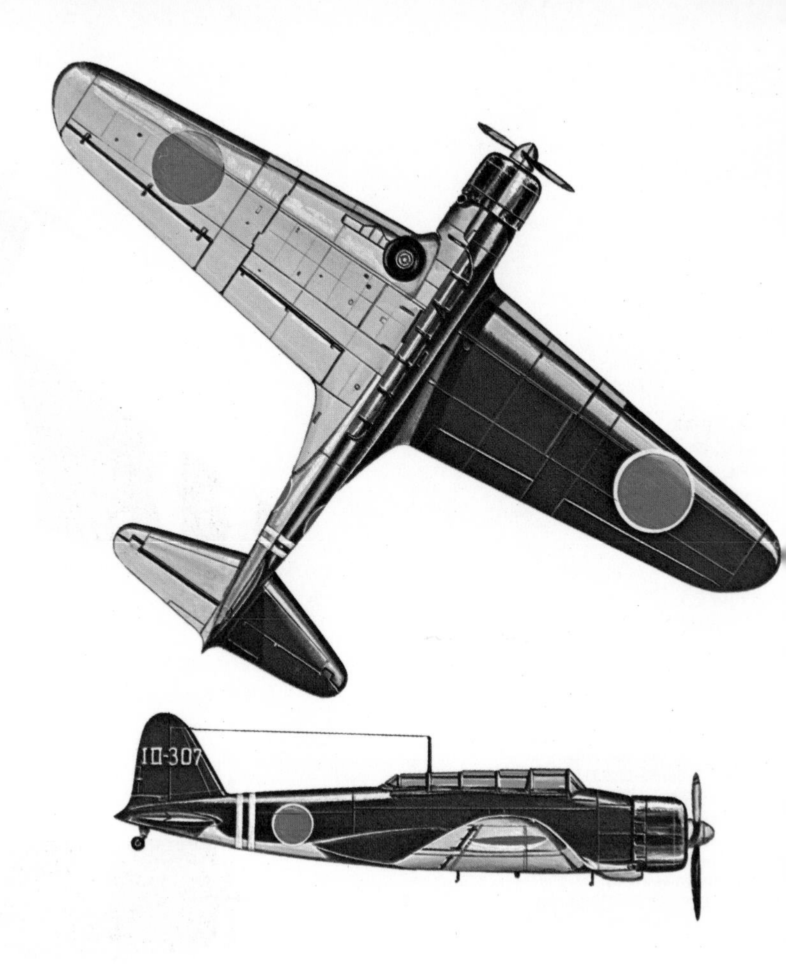

Nakajima B5N2 of the JNAF operating from the aircraft carrier *Zuikaku* during the Battle of the Coral Sea, May 1942

Span: 50ft 11in (15.52m)
Length: 33ft 9½in (10.30m)
Weight: 9,039lb (4,100kg)
Engine: 970hp Nakajima Sakae 11 14-cyl radial
Max speed: 235mph (378kmh) at 11,810ft (3,600ft)
Operational ceiling: 27,100ft (8,260m)
Range: 1,075 miles (1,730km)
Armament: 3 or 4×7.7mm machine-guns
Max bomb load: 1×1,764lb (800kg) torpedo, or 1,653lb (750kg) bombs, externally
externally

The B5N was designed by Nakajima to a 10-*Shi* (1935) requirement by the JNAF for a carrier-based attack bomber. It exhibited several advanced features for its time, and its design was influenced in no small part by the Northrop 5A, an example of which had arrived in Japan in 1935. The prototype B5N1 first flew in January 1937, and was a clean-looking 2/3-seat monoplane with mechanically folding wings, Fowler-type landing flaps and a fully retractable main undercarriage. Its 770hp Hikari 3 radial engine was installed beneath a neat NACA cowling, and on test the prototype aircraft exceeded the performance specification in many respects. The B5N1 Model 11 entered production late in 1937, and saw service in late spring/early summer 1938; it saw combat in the Sino Japanese conflict before Japan's entry into World War 2. The B5N1 was armed with a single defensive machine-gun at the rear. Production of the B5N1 was simplified by adopting manual instead of mechanical wing-folding, and by substituting slotted flaps for the Fowler type. Late production aircraft (designated Model 12) were fitted with 985hp Sakae 11 engines. A prototype appeared in December 1939 of the B5N2 Model 23, a torpedo bomber version powered by a 1,115hp Sakae 21 engine. This was armed with two forward-firing guns and one or two guns in the rear cockpit; and it could carry an 18in torpedo or bombs under the fuselage. The B5N2 entered JNAF service in 1940, and both variants were concerned in the attack on Pearl Harbor in December 1941. Subsequently they were responsible for the destruction of several other major US carriers during the early part of the war. The B5N, known by the Allied code name Kate, was encountered operationally as late as June 1944, when a number were engaged in the Marianas Islands campaign. For a year or more, however, the B5N had been obsolescent, and indeed several B5N1s were withdrawn from operations and converted to B5N1-K trainers soon after the B5N2 began to enter service. More than 1,200 Kates were completed, some of them by Aichi, before the B5N began to be superseded by its newer and faster stablemate, the B6N Tenzan. A number of B5Ns were, however, employed as suicide aircraft in the final stages of the war.

Short Sunderland

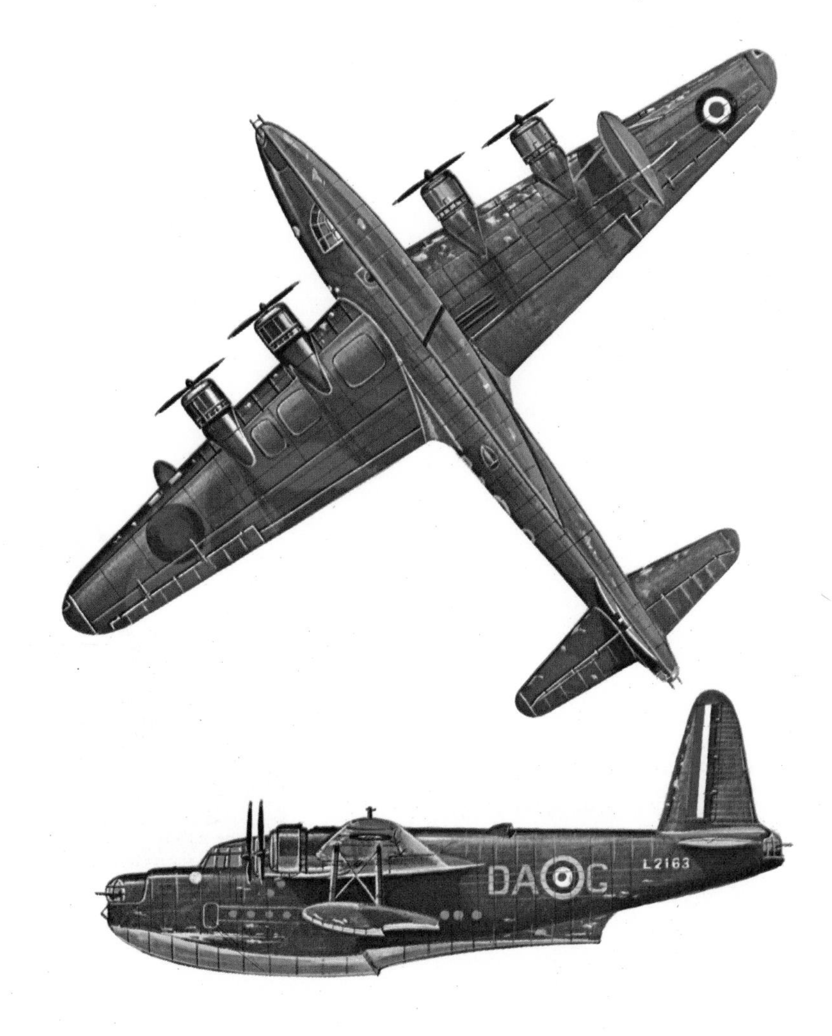

Short Sunderland Mk I of No 210 (GR) Sqdn, Coastal Command, RAF, 1940

Span: 112ft 8in (34.34m)
Length: 85ft 8in (26.11m)
Weight, normal take-off: 44,600lb (20,230kg)
Engines: 4×1,010hp Bristol Pegasus XXII 9-cyl radials
Max speed: 210mph (338kmh) at 6,500ft (1,980m)
Range: 1,780 miles (2,865km)
Armament: 4×0.303in Browning machine-guns; 3×0.303in Vickers K or Lewis machine-guns
Max bomb load: 2,000lb (907kg) internally

Design of the Sunderland, the RAF's longest-serving operational aircraft, was based on the successful C Class 'Empire' flying boats of Imperial Airways, and evolved to Specification R2/33. The prototype (K4774) flew on 16 October 1937 with 950hp Pegasus X engines. With Pegasus XXIIs and revised nose and tail armament, the Mk I entered production in 1938, and was first delivered to No 230 Sqdn in Singapore early in June 1938. By 3 September 1939 40 were in service with four RAF squadrons. Eventually 90 Mk Is were completed, 15 by Blackburn, which also built five of the 43 Mk IIs which, from late 1941, began to replace Mk Is in service. The Mk II introduced Pegasus XVIIIs, with two-stage superchargers, a twin-gun dorsal turret in place of beam gun ports, an improved tail turret and ASV radar. Rising operating weights necessitated redesign of the hull planing bottom, incorporated in the Mk III. The first Short-built Mk III flew on 15 December 1941; Shorts eventually produced 286; Blackburn built 170. No 10 Sqdn, RAAF first experimented with four nose machine-guns in the Mk III. This proved so successful against submarines and aircraft that many Sunderlands subsequently had this total armoury of ten guns, their bristling defence earning the respectful nickname porcupine from their German adversaries. The Mk IV was originally a larger, heavier development with 1,700hp Bristol Hercules, eight 0.50in machine-guns and two 20mm cannon. Only two prototypes and eight production aircraft were built, re-named Seaford, but after brief service were converted into Solent commercial transports for BOAC. Final military variant was the Mk V (100 built by Shorts, 50 by Blackburn), with 1,200hp Pratt & Whitney R-1830-90 Twin Wasps and improved ASV equipment. The Mk V entered service in February 1945, and was the last RAF version, finally retiring in 1958. Sunderlands exported postwar to the French *Aéronavale* (19) and the RNZAF (16) served until 1960 and 1966 respectively.

Vickers-Supermarine Spitfire

Supermarine Spitfire Mk I of No 19 Sqdn, RAF, Duxford, March 1939

Span: 36ft 10in (11.23m)
Length: 29ft 11in (9.12m)
Weight: 6,200lb (2,812kg)
Engine: 1,030hp Rolls-Royce Merlin II 12-cyl V-type
Max speed: 362mph (583kmh)
Operational ceiling: 31,900ft (9,725m)
Range: 395 miles (636km)
Armament: 8×0.303in Browning machine-guns

The Spitfire, Supermarine Type 300, was designed by Reginald Mitchell as a private venture using the new Rolls-Royce PV-12 Merlin. The Air Ministry later issued Specification F37/34. The prototype, K5054, flew on 5 March 1936, powered by a 990hp Merlin C. On 3 June 1936 the Air Ministry ordered 310. Deliveries began in August 1938. The Mk I was the principal model during the Battle of Britain; production totalled 1,583. Improvements included the 1,030hp Merlin II and III, three-blade propeller, and domed canopy. The suffix B later denoted wings carrying four machine-guns and two cannons, the original, standard eight-Browning installation becoming type A. The Mk II (1,175hp Merlin XII) entered service late in 1940; 920 were built, some later converted to Mk Vs. The Mk III (1,280hp Merlin XX) was experimental. The first really large-scale model, the Mk V, entered service in March 1941 (6,479 built). The C wing, introduced on the Mk VC, was a 'universal' wing mounting eight machine-guns, four cannon, or two cannon and four machine-guns. Clipped and pointed wing-tips were introduced for low and high altitude roles respectively. The pressurised Mk VI (100 built) and Mk VII (140) were high-altitude interceptors; the Mk VII had a redesigned fuselage and 1,710hp Merlin 64. The Mk VIII appeared in 1943 in high, intermediate and low-level forms, but was preceded in 1942 by the Mk IX (5,665 built), basically the Mk VC airframe with a Merlin 60-series. The Mk IX introduced the E wing with two Hispano and two 0.50in Brownings. The 16 Mk X built and 471 Mk XI were PR variants. A major stage was the Mk XII (100 built) of 1943, basically Mk VIII and IX airframes with a 1,735hp Griffon III or IV. The Mk XIV (957 built), a strengthened Mk VIII with a 2,050hp Griffon 65, was successful against the VI in 1944. The Mk XVI (1,054 built) appearing in 1944, was a Mk IX with a Packard-built Merlin 266. The unarmed PR Mk XIX (245 built), a Mk XIV derivative with a Griffon 65 or 66, served in Europe and the Far East. Postwar variants included the Mks XVIII, 21, 22 and 24. Production totalled 20,334, ending in October 1947. Spitfires of various marks were supplied during the war to the USAAF, the USSR, Egypt, Portugal and Turkey. The Seafire Mk IB (166 converted Spitfire Mk VBs) entered FAA service in mid-1942, followed by 372 Mk IICs, built as Seafires. The considerably improved Mk III of 1943 introduced folding wings. Postwar Seafires included the Mk XV (390 built), which introduced the Griffon, and the Mks XVII, 45, 46 and 47.

Potez 63

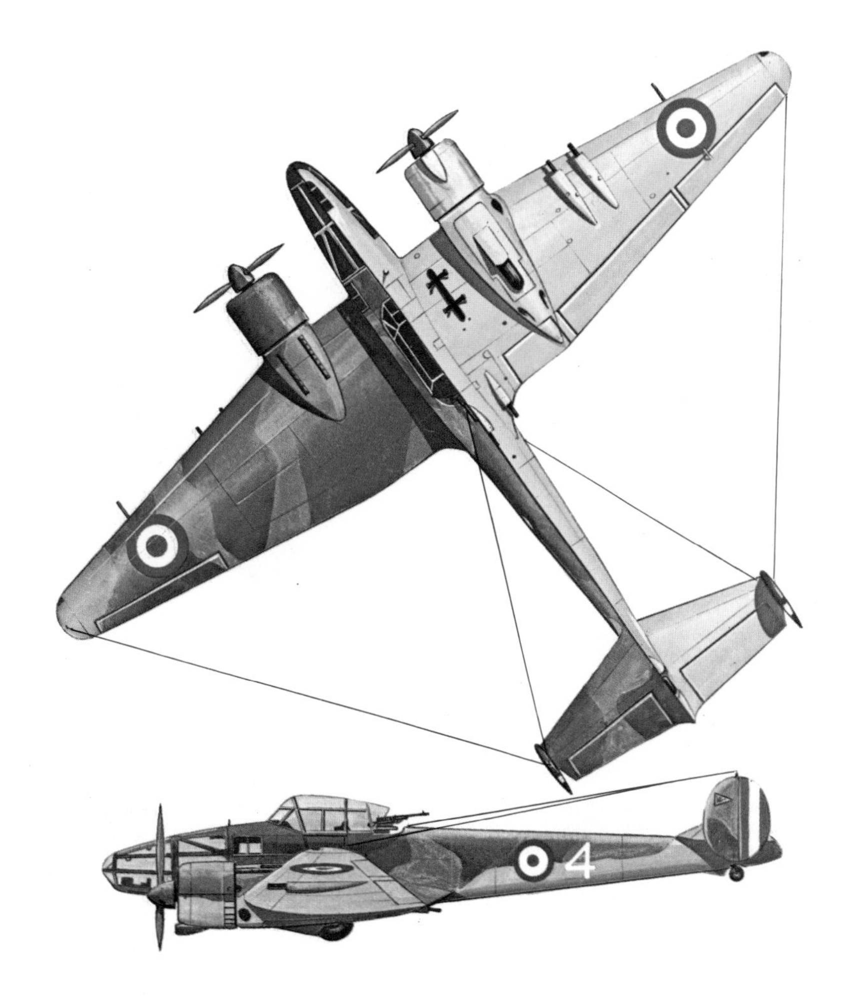

Potez 63-11 of *Groupe de Reconnaissance* II/39, *3ème Escadrille,* Vichy French Air Force, Syria 1941

Span: 52ft 5⅞in (16.00m)
Length: 36ft 1¼in (11.004m)
Weight, normal take-off: 9,773lb (4,433kg)
Engines: 2×700hp Gnome-Rhône 14m 14-cyl radials
Max speed: 264mph (425kmh) at 16,405ft (5,000m)
Operational ceiling: 29,530ft (9,000m)
Range: 932 miles (1,500km)
Armament: 8×7.5mm MAC 1934 machine-guns
Max bomb load: 176lb (80kg) internally; 220lb (100kg) externally

Developed to a 1934 specification for a 3-seat strategic fighter, the Potez 63 was designed by Coroller and Delaruelle, the -01 prototype first flying on 25 April 1936. In general appearance it bore a superficial resemblance to its German contemporary, the Messerschmitt Bf 110, a fact which contributed to the loss of many Potez fighters during combat in World War 2. The Hispano-Suiza-engined first prototype was later redesignated Potez 630-01, to distinguish it from the second machine (631-01), which had Gnome-Rhône engines. In May 1937 the French Air Ministry ordered ten evaluation machines which included representatives of both types, together with examples of the Potez 633 (light bomber configuration), 637 (reconnaissance and Army co-operation) and 639 (attack bomber). In June 1937 the Potez company became part of the new SNCA du Nord, from which was ordered 80 Potez 630 and 90 Potez 631, the latter figure including ten with dual controls for conversion training. Orders were placed in 1938 for 125 Potez 633 for the *Armée de l'Air,* plus export batches for Greece (24) and Romania (40); a manufacturing licence was also granted to Avia in Czechoslovakia, but none were built in that country. Only 11 of the Greek and 21 of the Romanian 633s had been delivered by August 1939, when delivery was halted by the French government. Production of the Potez 631 had by then reached 210 for the *Armée de l'Air* which accepted its first four in May 1938. Also completed were 60 Potez 637s, essentially an interim service model pending availability of the Potez 63.11, a much redesigned development which was first flown on 31 December 1938, differing principally in the design of the front fuselage; an initial order had been placed for 145 for the armed reconnaissance role. Nearly 1,700 more were ordered in 1939. Production continued under German direction after the occupation of France, and the eventual total of Potez 63.11s completed was in the region of 900. Variants of the Potez 63 series served with units of the *Luftwaffe* and Vichy Air Force, as well as with the FAFL, in Europe and North Africa.

Mitsubishi Ki-21 (Sally)

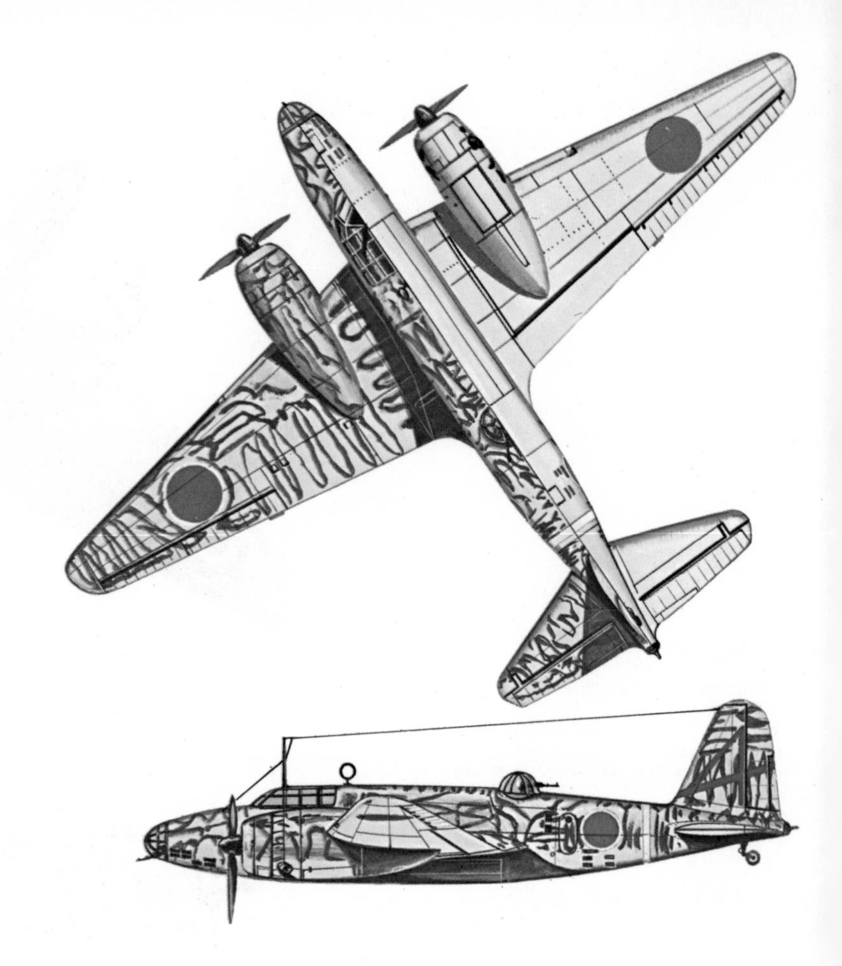

Mitsubishi Ki-21-IIb Model 2B of the 1st Sqdn, 14th Group, JAAF, Philippines, 1944

Span: 73ft 9⅞in (22.50m)
Length: 52ft 6in (16.00m)
Weight, normal take-off: 21,407lb (9,710kg)
Engines: 2×1,500hp Mitsubishi Ha-101 14-cyl radials
Max speed: 302mph (486kmh) at 15,485ft (4,720m)
Operational ceiling: 32,810ft (10,000m)
Range, normal: 1,678 miles (2,700km)
Armament: 5×7.7mm Type 89 machine-guns; 1×12.7mm Type 1 machine-gun
Max bomb load: 2,205lb (1,000kg) internally

The Ki-21 won an exacting design competition initiated by the Japanese Army Air Force early in 1936, and the first of five prototypes was completed in November of that year. With improved fields of fire for the defensive guns, and 850hp Nakajima Ha-5-*Kai* engines replacing the Mitsubishi Kinsei Ha-6s of the first prototype, it was accepted for initial production as the Ki-21-Ia, or Type 97 heavy bomber. It entered service in autumn 1938. In 1938 Nakajima also began production, delivering its first Ki-21 in August. The Model 1A was quickly succeeded by the Model 1B (Ki-21-Ib), into which were built modifications resulting from combat experience gained in China. Increases were made in protective armour for the crew, defensive armament and the sizes of the flaps and bomb bay. The Model 1C (Ki-21-Ic) had increased fuel and an extra lateral gun. A wider-span tailplane was introduced on the Model 2A (Ki-21-IIa). Mitsubishi began the development of this model late in 1939, introducing 1,490hp Ha-101 engines. The Ki-21 was a standard Army bomber at the time of Pearl Harbor, and was subsequently encountered in Burma, Hong Kong, India, Malaya, the Netherlands East Indies and the Philippines. Under the Allied code-naming system, the Ki-21 was known as Sally, although the name Gwen was briefly allocated to the Model 2B (Ki-21-IIb) before it was recognised as a Ki-21 variant. The Model 2B was the final production variant of this now-obsolescent bomber, recognisable by the turret replacing the dorsal 'greenhouse' of the earlier models. With its appearance, many earlier Ki-21s were withdrawn either for training or for conversion to MC-21 transports. Shortly before the war ended, nine Ki-21s were made ready as assault transports at Kyushu, to transfer demolition troops to Okinawa, but only one reached its target. Production came to an end in September 1944 after 1,713 had been built by Mitsubishi, plus 351 by Nakajima (up to February 1941). Just over 500 transport counterparts of the Ki-21 were built by Mitsubishi, designated MC-20 in their civil form and Ki-57 (code-name Topsy) in military guise. Proposals for a Ki-21-III version were shelved in favour of the Ki-67.

Vickers Wellington

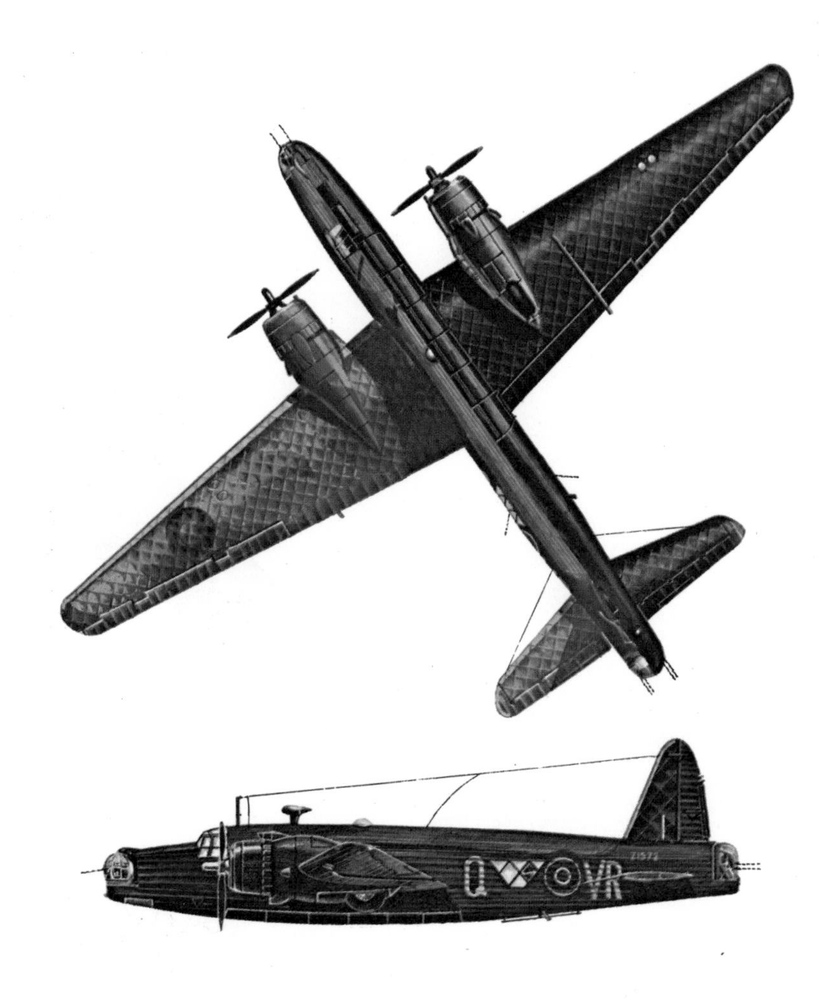

Vickers Wellington Mk III of No 419 Sqdn, RCAF, 1942

Span: 86ft 2in (26.26m)
Length: 64ft 7in (19.68m)
Weight: 29,500lb (13,381kg)
Engines: 2×1,500hp Bristol Hercules XI 14-cyl radials
Max speed: 255mph (410kmh) at 12,500ft (3,810m)
Operational ceiling: 19,000ft (5,790m)
Range with 1,500lb (680kg) bomb load: 2,200 miles (3,541km)
Armament: 8×0.303in machine-guns
Max bomb load: 4,500lb (2,041kg) internally

On 3 September 1939 the RAF had 179 Wellingtons on strength, rather less than the number of Hampdens or Whitleys, but these types bore the brunt of Bomber Command's operations in the early part of the war until the 4-engined heavy bombers arrived from 1941. The Wellington outstripped both of its contemporaries, a total of 11,461 being built before production ceased in October 1945. After being withdrawn from Bomber Command in 1943, Wellingtons served with Coastal Command as maritime reconnaissance aircraft, at home and in the Middle and Far East; others were employed briefly as transports and after the war as aircrew trainers. The 'Wimpey' was designed to Specification B9/32, the prototype (K4049) flying on 15 June 1936. Production was ordered to Specification 29/36, the first flying on 23 December 1937. The first Wellington squadron, No 9, received its aircraft in October 1938. Those in service when war broke out were Pegasus-engined Mks I or IA, but the most numerous early model was the Mk IC (2,685). Prototypes had flown before the war of the Merlin-engined Mk II and of the Mk III with Bristol Hercules radials. Wellingtons of Nos 9 and 149 Sqdns, in company with Blenheims, carried out the RAF's first bombing attack of the war, attacking German shipping at Brunsbüttel. From mid-December 1939 they were switched to night bombing only, joining in the first raid on Berlin in August 1940. In September they made their Middle East debut, and appeared in the Far East from early 1942. By this time the Mk III (1,519 built) was the principal service version, although two squadrons operated the Twin Wasp-engined Mk IV. The first general reconnaissance version for Coastal Command, the Mk VIII, appeared in spring 1942; 394, with similar engines to the Mk IC, were built. Substantial batches of the Mks XI, XII, XIII and XIV, with differing versions of the Hercules and variations in operational equipment, followed. Overseas, the Wellington maintained its combat role, the Mk X in particular (3,804 built) serving with the Middle East Air Forces as well as with Bomber Command. Wellingtons of No 40 Sqdn bombed Treviso, Italy, as late as March 1945. Wellingtons were converted as torpedo bombers, mine-layers and transports, and a special variant designated DW1 was fitted with a large electro-magnetic 'de-gaussing' ring to trigger off mines. The light but extremely strong geodetic construction of the Wellington not only enabled it to carry a creditable bomb load but withstood a considerable amount of battle damage.

Boeing B-17 Flying Fortress

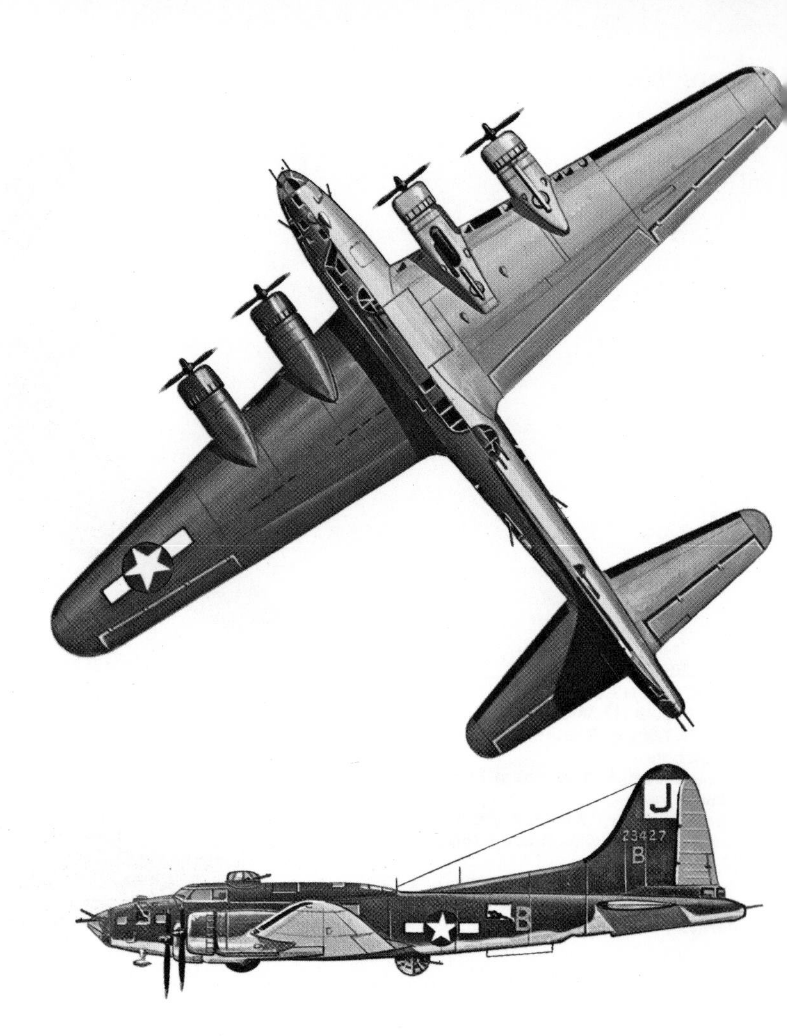

Douglas-built Boeing B-17F-60-DL of the 390th BG, US Eighth Air Force, UK, 1943-45

Span: 103ft 9⅜in (31.63m)
Length: 74ft 8⅞in (22.78m)
Weight: 55,000lb (24,948kg)
Engines: 4×1,200hp Wright R-1820-97 Cyclone 9-cyl radials
Max speed: 299mph (481kmh) at 25,000ft (7,620m)
Operational ceiling: 37,500ft (11,430m)
Range with 6,000lb (2,722kg) bomb load: 1,300 miles (2,092km)
Armament: 8×0.50in Browning machine-guns, 1×0.303in Browning machine-gun
Max bomb load (short range): 12,800lb (5,806kg) internally, plus 8,000lb (3,629kg) externally

In August 1934 Boeing began to develop their Model 299 bomber design to meet a 1934 USAAC requirement for an offshore anti-shipping bomber. The prototype, powered by four 750hp Pratt & Whitney radials, flew on 28 July 1935. Just over three weeks later it began USAAC evaluation trials but was destroyed on 30 October 1935 in a take-off accident. However, it had shown sufficient potential and in January Boeing received a contract for 13 YB-17 and one Y1B-17A for more exhaustive service trials, these differing from the prototype in having 930hp Wright Cyclones, those of the Y1B-17A being turbo-supercharged. Twelve Y1B-17s entered service with the Army's 2nd BG, with whom they remained as B-17s after completing service trials. In 1938 a production contract was placed for the modest quantity of 39. These incorporated a number of improvements, notably a modified nose and a larger rudder, and were designated B-17B. The 38 B-17Cs had increased armament and more powerful Cyclones; 20 were supplied to the RAF in 1941 as the Fortress Mk I; 42 B-17Ds had a tenth crew member, self-sealing fuel tanks and no external bomb racks. Most of the USAAF's remaining B-17Cs were converted to D standard. The last variant to appear before Pearl Harbor was the B-17E, which first flew in September 1941. It introduced the huge ventral fin and rudder that characterised subsequent Fortresses; firepower was considerably increased by ventral and dorsal power turrets, a tail gun position and a multi-gun nose, bringing defensive armament to 13 guns. Boeing built 512 B-17Es, including 45 RAF Fortress Mk IIs. B-17s carried out the first raids on European targets made by the 8th US Air Force in August 1942; this version also served extensively in the Pacific theatre. The B-17F was sub-contracted to Douglas and Lockheed-Vega factories which, with Boeing, built 3,405; 19 were supplied to the RAF as Fortress Mk IIs, and 41 were converted to F-9 PR aircraft. These companies built 8,680 of the last production model, the B-17G; 85 became Fortress Mk IIIs with RAF Coastal Command, and ten were converted to F-9Cs. The B-17G was characterised by its 'chin' turret with two 0.50in machine-guns, also added to later B-17Fs in service. The USN and USCG were allotted 48 B-17Gs for ASR or early warning patrol duties, designated PB-1G and PB-1W respectively. About 50 B-17s, adapted to carry a lifeboat under the fuselage, were redesignated B-17H and employed on ASR.

Messerschmitt Bf 110

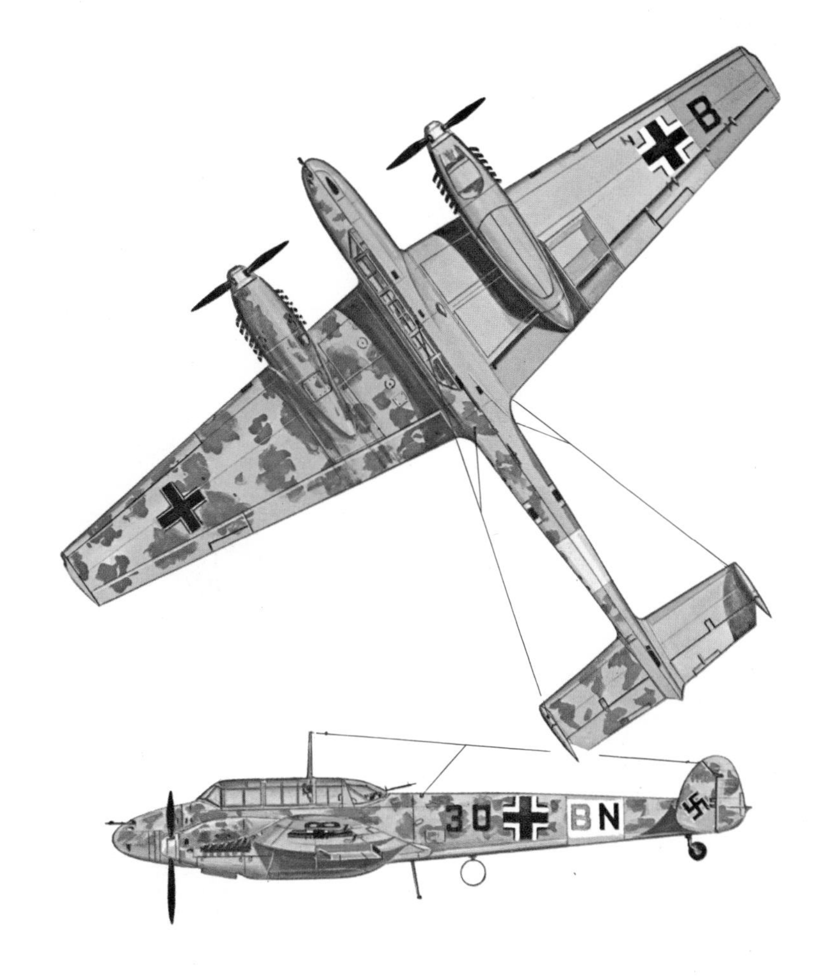

Messerschmitt Bf 110C-1 of 5/*ZG* 26, *Luftwaffe,* North Africa, 1942

Span: 53ft 3¾in (16.25m)
Length: 39ft 7¼in (12.07m)
Weight, normal take-off: 13,289lb (6,028kg)
Engines: 2×1,100hp Daimler-Benz DB601A-1 12-cyl inverted V-type
Max speed: 336mph (540kmh) at 19,685ft (6,000m)
Operational ceiling: 32,810ft (10,000m)
Range: 876 miles (1,410km)
Armament: 2×2mm MGFF cannon; 4×MG 17 machine-guns; 1×7.9mm MG15 machine-gun

The Bf 110 was the second production warplane designed by Prof Willy Messerschmitt for BFW AG. Developed to an RLM specification of early 1934 for a long-range escort fighter and *Zerstörer* (destroyer) aircraft, three prototypes were completed with DB600 engines, the first flying on 12 May 1936. The second, delivered early in 1937 to the *Luftwaffe* for service trials, was fast for a relatively heavy twin-engined machine, but was heavy on the controls and less manoeuvrable than desired. Four pre-series Bf 110A-0s were ordered which, due to the scarcity of DB600 engines, were fitted with 610hp Jumo 210Bs. These were inadequate, and were succeeded in spring 1938 by two Bf 110B-0s with 690hp DB600As to carry out trials for the initial Bf 110B-1 production series. Operational evaluation of the B-1 in the Spanish Civil War was forestalled when that conflict ended. Thus, the first model in active service was the Bf 110C, with increased power provided by DB601As, square wingtips to improve manoeuvrability, and a modified crew enclosure. Entering service in January 1939, over 500 were on strength by the year's end. Produced for fighter-bomber and reconnaissance roles, the Bf 110 was employed primarily for ground attack during the invasion of Poland and not until it was fully exposed as a fighter in the Battle of Britain were its shortcomings in that capacity revealed. Losses became so heavy that Bf 109s were sent with the bombers to protect their Bf 110 escorts. Production of the C continued, latterly with 1,200hp DB601Ns, but many earlier machines were withdrawn to second-line duties. Attempts to boost range produced the Bf 110D, a fighter (D-0 and D-1) and a fighter-bomber (D-2 and D-3). By mid-1941 most C and D versions were operational only in the Middle East or on the Eastern Front. The more versatile E (DB601N) and F (DB601F) appeared later that year, including the rocket-firing F-2 and F-4 night fighter. By late 1942, when it became apparent that the Me 210 was not a satisfactory replacement, Bf 110 production was stepped up again. The G was introduced, following the pattern of earlier series, including the G-4 night fighter with 1,475hp DB605B engines, two or four 20mm cannon and four 7.9mm machine-guns. The Bf110G-4/R3 introduced Lichtenstein SN-2 airborne interception radar. The H series, having heavier armament, was the last production model, produced in parallel with the G series. Total Bf 110 production was approximately 6,150, ending early in 1945.

Junkers Ju 88

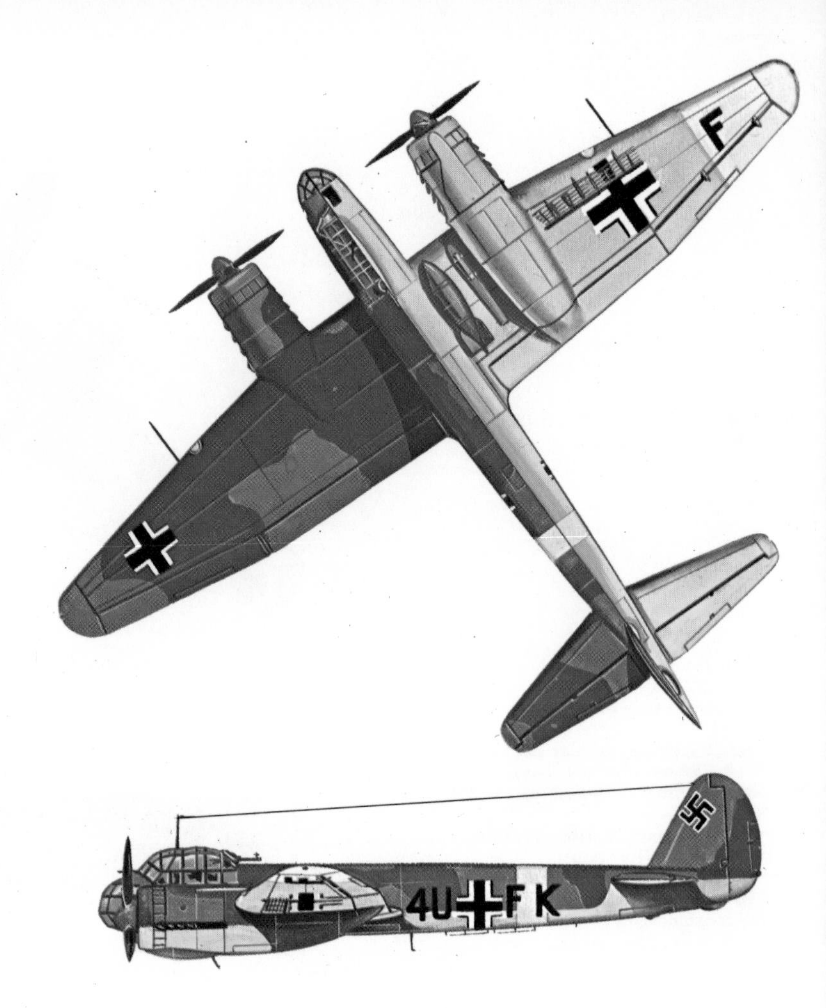

Junkers Ju 88A-4/Trop of 2*(F)*/123, *Luftwaffe*, Western Desert, summer 1942

Span: 65ft 7¾in (20.00m)
Length: 47ft 2⅞in (14.40m)
Weight: 30,865lb (14,000kg)
Engines: 2×1,340hp Junkers Jumo 2115 12-cyl inverted V-type
Max speed: 269mplh (433kmh) at 14,765ft (4,500m)
Operational ceiling: 26,900ft (8,200m)
Range: 1,112 miles (1,790km)
Armament: 7×7.9mm MG81 machine-guns
Max bomb load: 1,102lb (500kg) internally, 2,205lb (1,000kg) externally

The most adaptable German warplane of World War 2, and among the most widely used, the Ju 88 was developed to a 1935 RLM requirement for a high-speed bomber. The first prototype flew on 21 December 1936. The first and second prototypes had two 1,000hp DB 600A V-type engines, but the third had Jumo 211As and the Jumo powered the majority of Ju 88s. The fourth had the characteristic multi-panelled glazed nose. Following a pre-series batch of Ju 88A-0s, delivery of production Ju 88A-1s began in September 1939. The A series continued, with a few gaps, to the A-17, including dive bombing, anti-shipping strike, long range reconnaissance and conversion training variants. Probably the most common, the A-4 served in Europe and North Africa and was supplied to Finland and Italy. The first version with modifications gained from experience during the Battle of Britain, it had extended-span wings, Jumo 211Js and increased bomb load and armament. The Ju 88B was developed into the Ju 188. The Ju 88D (over 1,800 built as D-1, D-2 and D-3), was developed from the A-4 for strategic reconnaissance. The Ju 88S bomber – 1,700hp BMW 801G radials (S-1), 1,810hp BMW 801TJs (S-2) or 1,750hp Jumo 213E-1s (S-3) – differed from the earlier bombers in having a smaller, fully rounded glazed nose, and less armament and bomb load, but performance was considerably better. The Ju 88T-1 and T-3 were PR counterparts of the S-1 and S-3. Many Ju 88s were used as the explosive-laden lower portion of *Mistel* composite attack weapons. Paralled with the bomber series Junkers developed Ju 88 'heavy' fighters. The first was the Ju 88C-2, a Ju 88A-1 conversion with a 'solid' nose mounting three MG17 machine guns and a 20mm MGFF cannon, and an aft-firing MG15, entering service late in 1940. Small batches of the C-4, with the A-4's extended wings, and the C-5 followed. Final C sub-types were the C-6 day fighter (Jumo 211Js), with three nose cannon, and an MG131 replacing the MG15, and the C-7 night fighter. The Ju 88G night fighter, appearing from mid-1944, utilised the Ju 188's angular vertical tail and carried Lichtenstein radar. A small batch of H-2 'heavy' fighters was built. Final fighter variants were the Ju 88R-1 and R-2, day and night fighters respectively. The specialised ground attack/anti-tank Ju 88P, for the Russian front, had a 75mm (P-1) or two 37mm cannon in the nose. Ju 88 production totalled 14,676 (omitting prototypes) including 10,774 bomber and reconnaissance variants.

Dewoitine 520

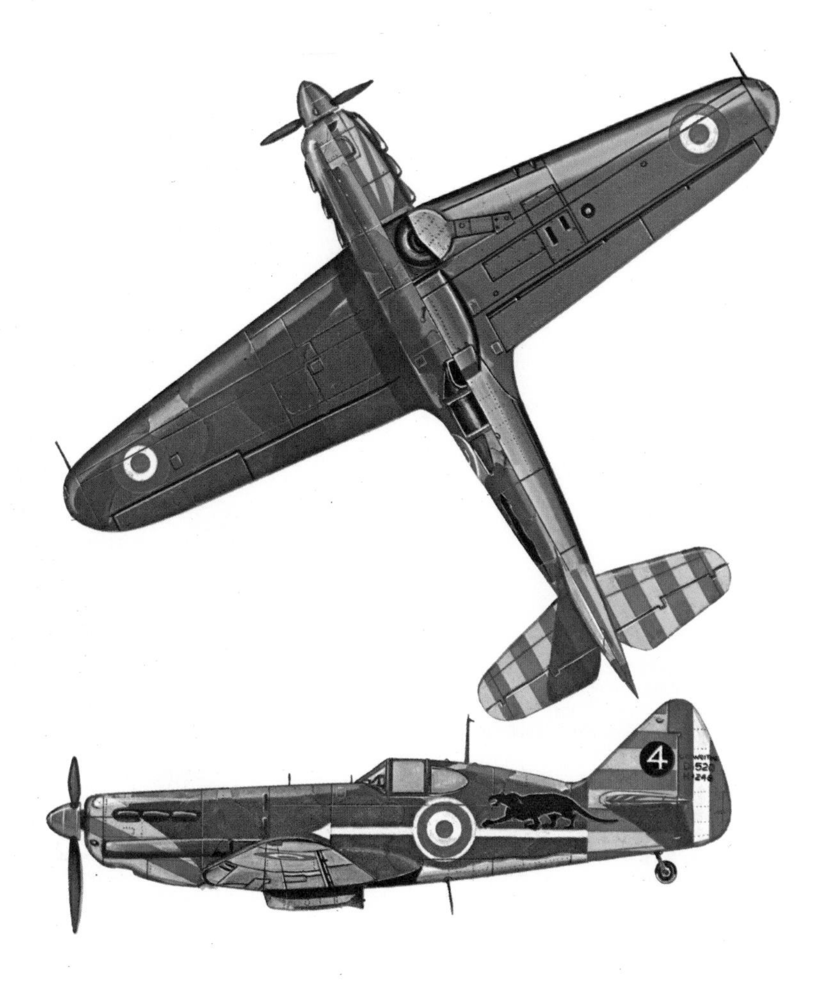

Dewoitine 520 of *Groupe de Chasse* 11/7 *4e Escadrille,* Vichy Air Force, Tunisia, early 1942

Span: 33ft 5⅝in (10.20m)
Length: 28ft 8⅞in (8.76m)
Weight: 6,135lb (2,783kg)
Engine: 930hp Hispano-Suiza 12Y-45 12-cyl V-type
Max speed: 326mph (525kmh) at 19,685ft (6,000m)
Operational ceiling: 36,090ft (11,000m)
Range: 615 miles (990km)
Armament: 1×20mm Hispano HS404 cannon; 2×7.5mm MAC 1934-M39 machine-guns

Design of the D 520, by Robert Castello, was initiated by Dewoitine in mid-1936 as a private venture. After some initial lack of enthusiasm, the French ordered two prototypes of a modified version in April 1938 from the SNCA du Midi, which had absorbed the Dewoitine company. The first of these was flown on 2 October 1938, powered by an 860hp Hispano-Suiza 12Y-21 engine, and in later trials with a 12Y-29 engine it attained its design speed of 373mph (520kmh). The second prototype carried armament and incorporated a number of structural and aerodynamic improvements, including redesigned tail surfaces. An initial order was placed in April 1939 for 200 D 520s, and successive orders (and cancellations) up to April 1940 required a total of 2,200 to be built for the *Armée de l'Air* and 120 for the *Aéronavale.* Production aircraft, with a slightly longer fuselage, increased fuel tankage and armour protection for the pilot, were powered by Hispano-Suiza 12Y-45 engines, and delivery began in January 1940 to an experimental flight at Bricy. When the German offensive in France began on 10 May 1940, only 36 D 520s were in service, with *Groupe de Chasse* I/3. These fought their first actions against the *Luftwaffe* on 13 May. In all, D 520s served with five *Groupes de Chasse* during the May-June fighting, destroying well over 100 enemy aircraft for a loss of 54 due to enemy action. After 25 June 1940, well over 300 D 520s (of 437 then built) survived either in unoccupied France or in North Africa, and the latter were utilised by four *Groupes* of the Vichy French Air Force and one *Escadrille* of the *Aéronavale.* In 1941 the German authorities ordered the production of 550 more D 520s: 349 were completed. In 1943-44, following the occupation of the remainder of France and the disbandment of the Vichy Air Force, the SNCA du Sud-Est completed a further quantity for German use, bringing overall production of the D 520 to 905 aircraft. In addition to the *Luftwaffe,* the air forces of Bulgaria, Italy and Romania were also supplied with quantities of the French fighter. Aircraft recaptured by the Allies, as France was progressively liberated, fought with the *Forces Françaises de l'Interieur* during the final months of the war in Europe.

Douglas SBD Dauntless

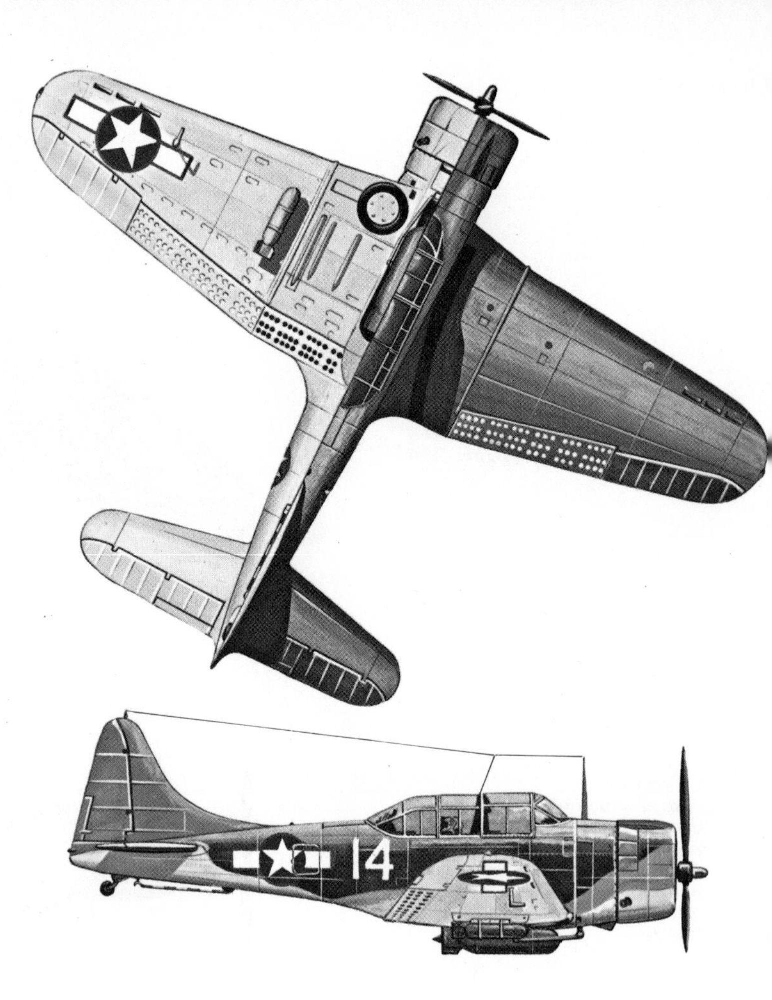

Douglas SBD-5 Dauntless of VB-5, US Navy, USS *Yorktown*, August/September 1943

Span: 41ft 6in (12.65m)
Length: 33ft 0in (10.06m)
Weight: 10,700lb (4,853kg)
Engine: 1,200hp Wright R-1820-60 Cyclone 9-cyl radial
Max speed: 252mph (406kmh) at 13,800ft (4,200m)
Operational ceiling: 24,300ft (7,400m)
Range: 1,115 miles (1,794km)
Armament: 2×0.50in machine-guns; 2×0.30in machine-guns
Max bomb load: 1,000lb (454kg) externally

Evolution of the Dauntless began in 1934, when a Northrop team under Ed Heinemann based a Navy dive-bomber on Northrop's Army A-17A. Designated XBT-1, it flew in July 1935. In February 1936 54 BT-1s with 825hp R-1535-94 engines were ordered. The last was completed as the XBT-2, with a 1,000hp R-1820-32 engine; with further modifications, it was redesignated XSBD-1 when Northrop was absorbed by Douglas on 31 August 1937. Perforated dive flaps, a distinctive Dauntless feature, were then introduced. Delivery of 57 SBD-1s to the USMC began in June 1940. Simultaneously, the USN ordered 87 SBD-2s with additional fuel and armour, and autopilots. Both versions had two 0.30in machine-guns in the upper cowling and a 0.30in in the rear cockpit. Bombs up to 1,000lb (454kg) could be carried on a ventral cradle; maximum bomb load was 1,200lb (544kg). Delivery of SBD-2s, from November 1940, was followed from March 1941 by 174 -3s with R-1820-52 engines and 0.50in front guns. The two models were standard USN carrier-borne dive-bombers at the time of Pearl Harbor; subsequently, the Navy received a further 410 SBD-3s. In May 1942 SBD pilots were credited with 40 of the 91 enemy aircraft lost during the Battle of the Coral Sea; at Midway SBDs sank three Japanese carriers and crippled one. Their own attrition rate was the lowest of any US carrier aircraft in the Pacific, due largely to an outstanding ability to absorb battle damage. Later, Dauntlesses operated from escort carriers, flying ASW or close-support missions. In October 1942 delivery began of 780 SBD-4s with radar and radio-navigation equipment. The major production model, the SBD-5, with increased engine power, followed. In addition to the 2,965 SBD-5s, 60 SBD-5As were delivered to the USMC: 450 SBD-6s completed Dauntless production in July 1944. Production totalled 5,936, including 168 A-24s and 615 A-24Bs for the USAAF, delivered from June 1941, these corresponding to the SBD-3 and -3A, -4 and -5 respectively, but had new tailwheels, internal equipment changes and no arrester gear. They were not flown with great combat success, and were used chiefly for training or communications. The RNZAF received 18 SBD-3s, 27 SBD-4s and 23 SBD-5s; 32 SBD-5s were supplied to the French Navy, and between 40 and 50 A-24Bs to the *Armée de l'Air;* but the latter, like their US Army counterparts, were employed mainly on second-line duties. Nine SBD-5s were delivered to the FAA but were not used operationally.

Grumman F4F Wildcat

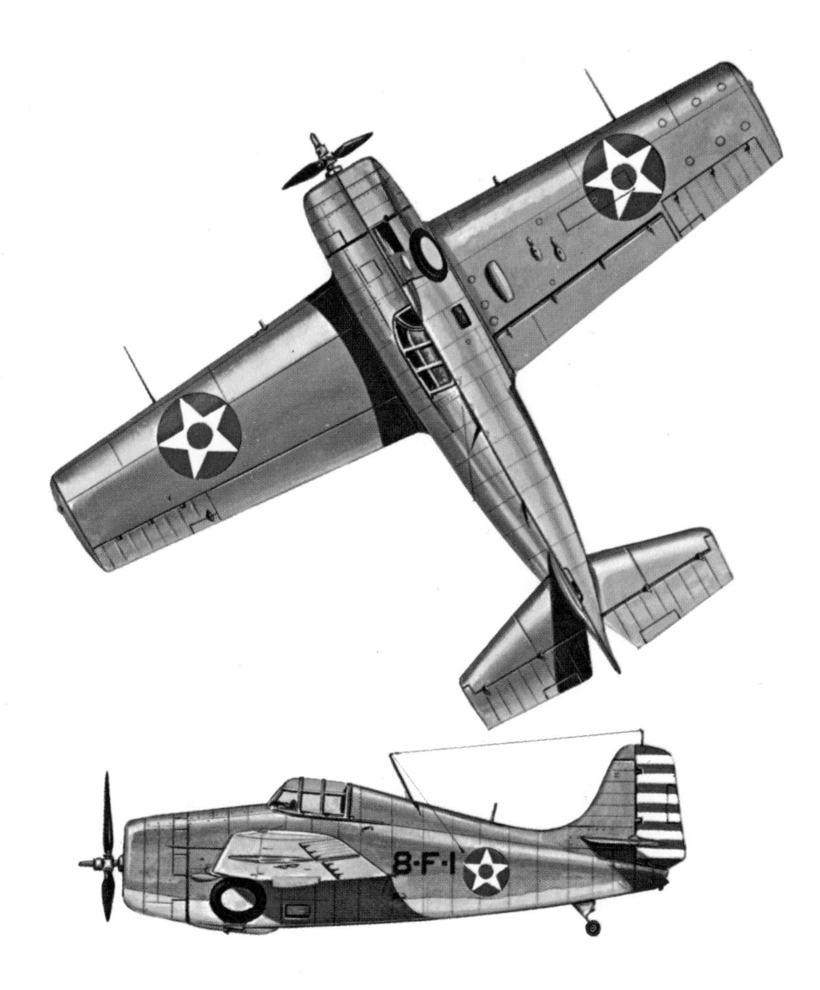

Grumman F4F-3 Wildcat of VF-8, US Navy, USS *Hornet,* late 1941

Span: 38ft 0in (11.58m)
Length: 28ft 9in (8.76m)
Weight, normal take-off: 7,002lb (3,176kg)
Engine: 1,200hp Pratt & Whitney R-1830-76 Twin Wasp 14-cyl radial
Max speed: 330mph (531kmh)
Operational ceiling: 37,500ft (11,430m)
Range, normal: 845 miles (1,360km)
Armament: 4×0.50in M-Z Browning machine-guns
Max bomb-load: 200lbs (96kg)

Originally, the Grumman proposals which won a 1936 USN development contract were for a biplane carrier fighter based on its earlier successful biplane types but this design, the XF4F-1, was shelved in favour of a monoplane fighter. The prototype, the XF4F-2, was flown on 2 September 1937 powered by a 1,050hp R-1830-66 Twin Wasp engine. The Navy decided to develop this still further, by ordering it to be rebuilt in a much-redesigned form as the XF4F-3, with an improved, supercharged XR-1830-76 engine. This aircraft flew on 12 February 1939, and was followed six months later by an initial production order for the F4F-3. Eventually, 285 F4F-3s were built. Deliveries to the USN late in 1940 were preceded by an order from France for 100 G-36A fighters, the export designation of the F4F-3 when fitted with a 1,200hp Wright R-1820-G205A engine. This order, later reduced to 81, was diverted to Britain in mid-1940 after the fall of France, these aircraft and nine others being employed by the FAA under the title Martlet Mk I; 30 G-36As, ordered by Greece, were also diverted to Britain to become Martlet Mk IIIs. Neither the F4F-3 nor the Martlet Mk I had wing-folding, but this feature was incorporated in all but the first ten of an order for 100 Martlet Mk II (G-36B) fighters placed by Britain in 1940. (The other ten corresponded to the USN's 65 F4F-3As, having non-folding wings and R-1830-90 engines). The USN's first folding-wing Wildcat was the Twin Wasp-engined F4F-4, Grumman building 1,389, including 220 F4F-4Bs with Cyclone engines as Martlet Mk IVs for the FAA. The Eastern Aircraft Division of General Motors delivered 838 similar but four-gunned aircraft, designated FM-1, to the USN and 222 to the FAA as Martlet Mk Vs. Eastern also built the FM-2, production version of Grumman's XF4F-8, with a 1,200hp Wright R-1820-56 Cyclone engine and taller fin and rudder. The USN received 4,407 and the FAA 370; the latter's were designated Wildcat Mk VI, the FAA having by now adopted the US name. Grumman's final production version (21 built) was the F4F-7, a heavier and slower unarmed version, with fixed wings, extra fuel and photo-reconnaissance cameras.

Mitsubishi A6M Zero-Sen (Zero/Zeke)

Mitsubishi A6M2 Model 21 Zero-Sen of the JNAF from the aircraft carrier *Soryu,* engaged in the attack on Port Darwin, Australia, February 1942

Span: 39ft 4½in (12.00m)
Length: 29ft 8¾in (9.06m)
Weight, normal take-off: 5,313lb (2,410kg)
Engine: 940hp Nakajima Sakae 12 14-cyl radial
Max speed: 332mph (535kmh) at 14,930ft (4,550m)
Operational ceiling: 32,810ft (10,000m)
Range, normal: 1,162 miles (1,870km)
Armament: 2×7.7mm Type 97 machine-guns; 2×20mm Type 99 cannon
Max bomb load: 264lbs (120kg)

Jiro Horikoshi designed the celebrated Zero fighter to the exacting 12-*Shi* (1937) JNAF specification for a carrier-borne fighter. It had the most widespread service career ever enjoyed by a Japanese combat aircraft. Two A6M1 prototypes were built, with 780hp Zuisei 13 radial engines. The first flew on 1 April 1939. Production began in 1940 with the A6M2 Model 11, the only major change being the adoption of the more powerful Sakae 12 engine. Operational evaluation of 15 Zeros in China followed, the JNAF officially accepting the type in July 1940. After 64 Model 11s were completed, the Model 21 with folding wingtips entered production in November 1940; it was the major version at the time of Pearl Harbor, although in mid-1941 the A6M3 Model 32 had appeared. Similar to the A6M2, except for its 1,300hp supercharged Sakae 21 engine, the A6M3's performance was later improved by removing the foldable wingtips, but this reduced manoeuvrability, and the full-span non-folding wing was restored in the A6M3 Model 22. In the fighting over Guadalcanal early in 1943 the Zero was no longer maintaining its superiority over its opponents. Hence the A6M5 Model 52 was developed, retaining the Sakae 21 but having a shorter span wing with square tips. Sub-types produced included the A6M5a Model 52A (strengthened wing and increased ammunition), A6M5b Model 52B (increased armament and armour), and A6M5c Model 52C (further armour; two 20mm and three 13mm guns), all appearing in 1944. The Model 52C's higher gross weight penalised performance and few were built. The A6M6c Model 53C had a Sakae 31 with methanol injection, bullet-proof fuel tanks and underwing rocket rails. When supplies of Sakae engines were compromised by Allied air attacks, the A6M8c Model 54C appeared with a 1,500hp Mitsubishi Kinsei 62, armed only with four wing guns. In 1945 Mitsubishi built 465 *Kamikaze* versions (A6M7 Model 63); several hundred other Zeros were expended in suicide attacks. A total of 10,937 Zeros was built by VJ-day. Mitsubishi built 3,879 of these, but the principal manufacturer was Nakajima, which produced 6,217 plus 327 twin-float A6M2-Ns. In addition, 508 A6M2-K and six A6M5-K 2-seat conversion trainers were built.

Bristol Beaufighter

Bristol Beaufighter TF Mk VIC of Coastal Command Development Unit, RAF, *circa* March/April 1943

Span: 57ft 10in (17.63m)
Length: 41ft 4in (12.60m)
Weight: 23,884lbs (10,834kg)
Engines: 2 x 1,670hp Bristol Hercules VI or XVI 14-cyl radials
Max speed: 312mph (502kmh) at 14,000ft (4,265m)
Operational ceiling: 26,000ft (7,925m)
Range: 1,540 miles (2,478km)
Armament: 4 x 20mm Hispano cannon; 6 x 0.303in Browning machine-gun; one 0.303in Vickers K machine-gun
Max bomb load: 1 x 1,605lb (728kg) torpedo

The Beaufighter, originally conceived as a fighter, originated in 1938 as a private venture design, based upon the Beaufort torpedo bomber. Specification F17/39 was issued to cover an initial order for 300 and the prototype, R2052, first flew on 17 July 1939. The first small batch of Mk IF production aircraft was accepted by the RAF in August 1940. With 10-gun armament and airborne interception radar, the early Beaufighters were the most potent night fighters in service; by the end of 1940 they were also operating as day fighters in the Western Desert. The Mk IC, a coastal protection and anti-shipping version similar to the Mk IF, and also powered by 1,590hp Hercules XI engines, appeared in early 1941. A total of 914 Mk Is was built. To avoid strain on Hercules engine output, the Merlin-engined Mk II, was ordered, and prototypes flew in 1940: 450 production Mk IIFs were built, with 1,280hp Merlin XXs. Most served as night fighters in the UK, but some were delivered to the FAA. The tendency towards instability in the Mk I increased in the Mk II with its longer nacelles, and was cured by giving the tailplanes 12 degrees of dihedral, a modification which became standard. The Mks III, IV and V were experimental variants. The next large-scale model was the Mk VI (830 built), produced, like the Mk I, as the Mk VIF and Mk VIC for Fighter and Coastal Commands respectively. It returned to a more powerful Hercules engine, enhancing performance and allowing a small bomb load to be carried, and introduced the observer's dorsal 0.303in Vickers K machine-gun. Mk VIFs served with the USAAF in 1943; others had an AI radar nose 'thimble'. The first torpedo-dropping experiments were made in 1942 with X8065, a Mk VC. The 'Torbeau' proved an effective torpedo bomber, but had enough performance to carry out its other coastal duties of escort fighter and reconnaissance. Sixty Mk VIs were completed as ITF (Interim Torpedo Fighters), before the TF Mk X torpedo-bomber and the non-torpedo-carrying Mk IXC, both with 1,770hp Hercules XVII engines, appeared. The TF Mk X (2,205 built), had the AI radar nose 'thimble', and a dorsal fin extension on later production batches. It could carry a heavier torpedo than the Mk VI (ITF) or 2,250lb (113kg) bombs and 8 rocket projectiles underwing. The Mk VI (ITF) aircraft were later brought up to Mk X standard. The Mk 21, built in Australia for the RAAF, was generally similar to the RAF's Mk X, except for Hercules XVIIIs and four 0.50in wing guns in place of six 0.303in Brownings. The Mk XIC was an interim model; only 163 were built. British production of all variants totalled 5,562, ending in September 1945.

Aichi D3A (Val)

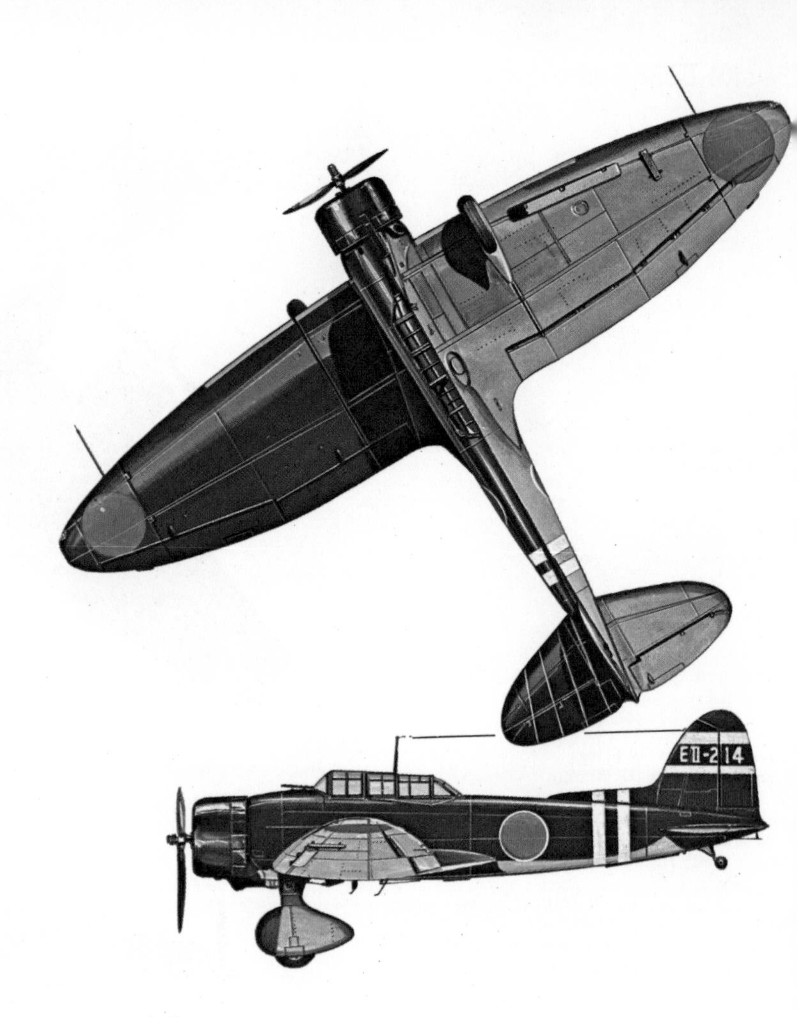

Aichi D3A1 Model 11 of the JNAF from the aircraft carrier *Zuikaku,* southern Pacific, 1941-42

Span: 47ft 1½in (14.365m)
Length: 33ft 5⅜in (10.195m)
Weight, normal take-off: 8,047lb (3,650kg)
Engine: 1,075hp Mitsubishi Kinsei 44 14-cyl radial
Max speed: 242mph (389kmh) at 7,610ft (2,320m)
Operational ceiling: 31,170ft (9,500m)
Range, normal: 1,131 miles (1,820km)
Armament: 3×7.7mm machine-guns
Max bomb load: 810lb (370kg) externally

Aichi was one of three companies to compete, in 1936, for the Imperial Japanese Navy's 11-*Shi* requirement for a new carrier-borne dive bomber, and its design showed strongly the influence of current Heinkel products, with whom the IJN had a clandestine agreement. Aichi's design was awarded a development contract, and entered service in October 1940 as the D3A1 Model 11, or Type 99 carrier-based dive bomber. The Model 11 remained in production until August 1942; it was a standard JNAF type at the time of the attack on Pearl Harbor, in which it took part, and in April 1942 the British carrier HMS *Hermes* and the cruisers *Cornwall* and *Dorsetshire* were sunk in the Indian Ocean by D3A1s. A single 250kg bomb could be carried on a ventral cradle which was swung forward and downward to clear the propeller during delivery, and a 60kg bomb could be attached to each outer wing section. After delivering its bombs the D3A1 was sufficiently well armed and manoeuvrable to put up a creditable fight against the Allied fighters then in service. A total of 478 D3A1s was built, production then continuing with the D3A2 Model 22 until January 1944, when 816 D3A2s had been completed. The D3A2 introduced cockpit and minor airframe improvements, but differed chiefly in having a 1,300hp Kinsei 54 engine which raised the maximum speed to 266mph (428kmh) at 18,540ft (5,650m). Normal and maximum take-off weights of this model were 8,378lb (3,800kg) and 9,088lb (4,122kg) respectively, and the maximum range 1,572 miles (2,530km). Both the D3A1 and the D3A2 (which were code named Val by the Allies) figured prominently in the major Pacific battles, including those of Santa Cruz, Midway and the Solomon Islands; but increasing losses, both of aircraft and of experienced pilots, progressively reduced their contribution to the Japanese war effort, and during the second half of the Pacific War they were encountered much less often. Some were converted as single-seat suicide attack aircraft, and a number of D3A2s were adapted for the training role, designated D3A2-K.

Handley Page Halifax

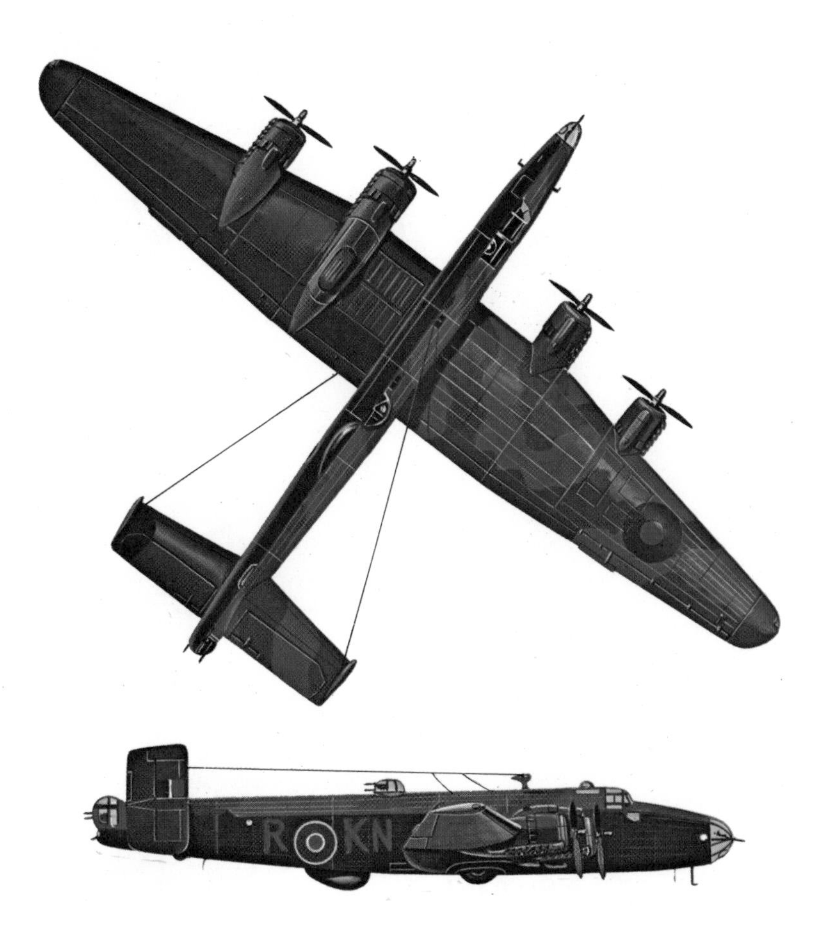

Handley Page Halifax B Mk III of No 77 Sqdn, Bomber Command, RAF, 1944

Span: 104ft 2in (31.75m)
Length: 71ft 7in (21.82m)
Weight: 54,400lb (24,675kg)
Engines: 4×1,615hp Bristol Hercules XVI 14-cyl radials
Max speed: 282mph (454kmh) at 13,500ft (4,115m)
Operational ceiling: 24,000ft (7,315m)
Range, with max bomb load: 1,030 miles (1,658km)
Armament: 8×0.303in Browning machine-guns; 1×0.303in Vickers K machine-gun
Max bomb load: 13,000lb (5,897kg)

The HP57 came into being as a redesign of the twin-Vulture-engined HP56 developed to meet Air Ministry Specification P13/36, after it had become apparent that the Vulture would be unsatisfactory. The HP57 was a much-enlarged design, drawn up around four Rolls-Royce Merlin engines, and the first of two prototypes (L7244) was flown on 24 September 1939, by which time an initial contract had been placed. Deliveries of the Halifax Mk I, to No 35 Sqdn, began in November 1940, and the bomber made its operational debut in a raid on Le Havre on the night of 11-12 March 1941. Early machines were known as Mk I Srs I, followed by the Srs II with a higher gross weight and Srs III with increased fuel tankage. The first major modification appeared in the Mk II Srs I, with its 2-gun dorsal turret and uprated Merlin XX engines. The Mk II Srs I (Special) had a fairing in place of the nose turret, and the engine exhaust muffs omitted; the Mk II Srs IA introduced the drag-reducing moulded Perspex nose that became a standard feature, had a 4-gun dorsal turret, and Merlin 22 engines. Variants of the Mk II Srs I (Special) and Srs IA, with Dowty landing gear instead of the standard Messier gear, were designated Mk V Srs I (Special) and Srs IA. A total of 1,966 Mk IIs and 915 Mk Vs was built. The Mk II Srs IA introduced larger, rectangular vertical tail surfaces, to overcome serious control difficulties experienced with earlier models. The Perspex nose and rectangular fins characterised all subsequent Halifaxes, whose only serious drawback now was a lack of adequate power. Thus, in the Mk III, which appeared in 1943, the Merlin powerplant was abandoned in favour of Bristol Hercules XVI radial engines. The Mk III became the most numerous variant, 2,091 being built. The Mk IV was an uncompleted project, the next operational models being the Mks VI and VII, the former powered by 1,675hp Hercules 100, the latter reverting to the Hercules XVI. These were the final bomber versions, and compared with earlier models were built in relatively small numbers. Halifax production ended with the Mk VIII supply transport and Mk IX paratroop transport, the final aircraft being delivered in November 1946, by which time a widely sub-contracted wartime programme had built 6,176 Halifaxes. The Halifax made its last operational sortie on 25 April 1945. It undertook special-duty missions, and it also served in the Middle East. After the war Halifaxes served for a time with Coastal and Transport Commands, the last flight being made by a Coastal Command GR Mk VI in March 1952.

North American B-25 Mitchell

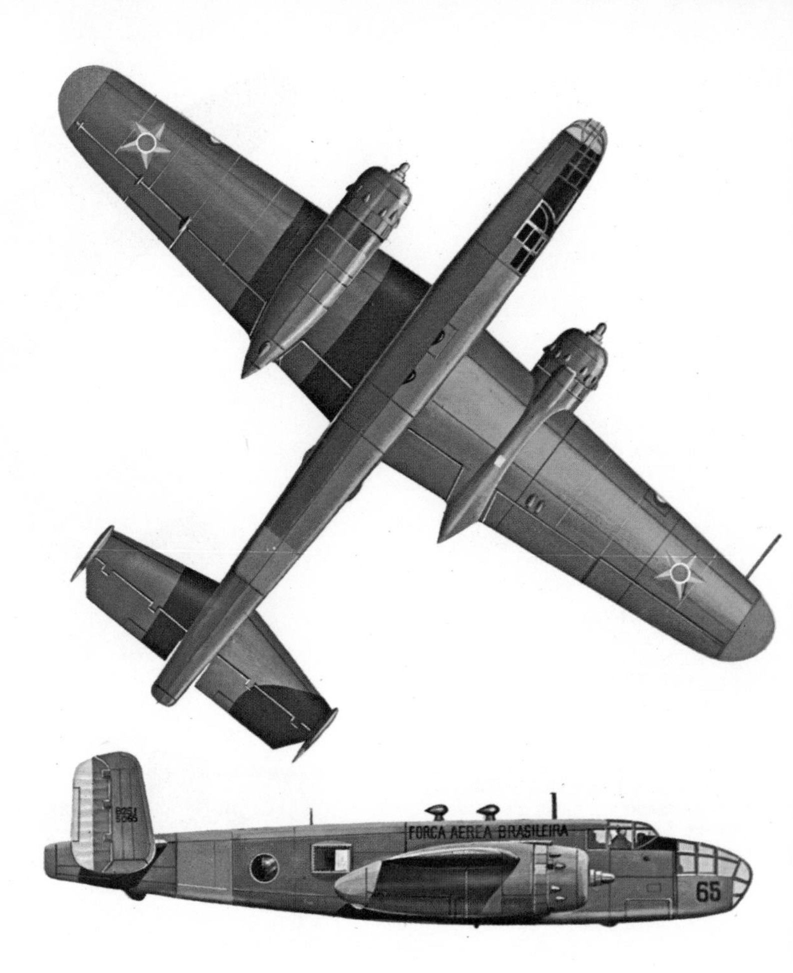

North American B-25J Mitchell of the Brazilian Air Force, *circa* 1960

Span: 67ft 7in (20.60m)
Length: 52ft 11in (16.13m)
Weight, normal take-off: 33,500lb (15,195kg)
Engines: 2×1,700hp Wright R-2600-29 14-cyl radials
Max speed: 275mph (433kmh) at 15,000ft (4,570m)
Operational ceiling: 24,200ft (7,375m)
Range, normal: 1,350 miles (2,173km)
Armament: 12×0.50in machine-guns
Max bomb load: 3,000lb (1,360kg) internally; 1,000lb (454kg) or 8×5in RPs externally (underwing)

North American Aviation was awarded an immediate production contract for its NA-62 design, without prototypes, and the first B-25 was flown on 19 August 1940. By the end of the year 24 had been delivered, having from the tenth the characteristic gull-wing. They were followed in 1941 by 40 B-25As, then by 119 B-25Bs with dorsal and ventral turrets. First operational unit was the 17th BG, which received B-25s in 1941. Production continued with 1,619 B-25Cs and 2,290 B-25Ds from early 1942. In April 1942 B-25Bs flew off the USS *Hornet* on their epic raid on Tokyo. The B-25E and F were experimental models. The B-25G carried two 0.50in guns alongside a 75mm cannon in a 'solid' nose. An even more heavily armed 'gunship' was the B-25H (1,000 built) with a 75mm nose cannon and 14 0.50in guns; this entered operational service in February 1944, joining the earlier multi-gunned Mitchells on anti-shipping strikes in the Pacific; 175 earlier B-25s had been modified to carry ten 0.50in guns. Final production model, the J (4,318 built), retained the forward placing of the dorsal turret introduced on the H. The transparent-nosed version of the J, most numerous variant of this adaptable aircraft, had five fixed and seven movable 0.50in machine-guns, while the solid nose J had 11 fixed guns, making a total of 18. The crew comprised five or six according to the number of gunners required. B-25s in US service operated predominantly in the Pacific, but large numbers were supplied elsewhere during the war. From 1943 248 B-25Hs and 458 B-25Js were transferred to the USN, as PBJ-1Hs and -1Js, most being operated by USMC squadrons. Under Lend-Lease 870 B-25s were supplied to the USSR, and 23 Mitchell Mk Is (B-25Bs) and 538 Mk IIs (B-25Cs and Ds) were received by the RAF. Others were supplied to Brazil (29), China (131) and the Netherlands (249). In the early postwar years a substantial number of B-25Ds were converted to target-tugs, and over 600 B-25Js became TB-25Js as aircrew trainers with the USAAF/USAF and RCAF, serving for many years. Until the 1960s a few B-25Js formed part of the bomber force of such air forces as those of Brazil, Uruguay and Venezuela.

Mitsubishi G4M (Betty)

Mitsubishi G4M2a Model 24 of the 753rd Air Corps, JNAF, Philippines, autumn 1944

Span: 81ft 7⅞in (24.89m)
Length: 64ft 4⅞ (19.63m
Weight: 33,069lb (15,000kg)
Engines: 2×1,850hp Mitsubishi Kasei 25 14-cyl radials
Max speed: 272mph (437kmh) at 15,090ft (4,600m)
Operational ceiling: 29,365ft (8,950m)
Range: 1,497 miles (2,410km)
Armament: 4×20mm Type 99 cannon; 1×7.7mm Type 97 machine-gun
Max bomb load: 2,205lb (1,000kg), of bombs, or 1×1,764lb (800kg) torpedo internally

The G4M evolved to a 'range-at-all-costs' specification, issued by the IJN in 1937 for a twin-engined medium bomber. Kiro Honjo, who led Mitsubishi's design team, could only accomplish this by packing so much fuel into the wings that no armour protection could be provided for the fuel tanks or the crew. After the prototype G4M1 had flown in October 1939, Mitsubishi was instructed to adapt the design for bomber escort duties, with increased armament and a crew of ten, designated G6M1. Thirty were built and service tested before the JNAF admitted their performance was inadequate and abandoned the project. They subsequently served as G6M1-K trainers and later as G6M1-L2 troop transports. By the end of March 1941, 14 more G4M1 bombers had flown, and in April this version was accepted for JNAF service as the Model 11 land-based attack bomber. The G4M1 had a single 20mm tail gun, with a 7.7mm gun in the nose, ventral and dorsal positions; powerplant was two 1,530hp Mitsubishi Kasei 11 radials. The G4M1 was used in pre-war operations in south-east China, and by the time Japan entered World War 2 there were some 180 G4M1s in JNAF service. They scored a number of early successes, but their 1,100 Imp galls (5,0001) of unprotected fuel made them extremely vulnerable and they soon became known to US gunners as the 'one-shot lighter.' Three months after particularly heavy losses in the Solomons campaign of August 1942, Mitsubishi began work on the G4M2 Model 22. This had an improved armament, and 1,850hp Kasei 21 engines with methanol injection; fuel capacity was increased, but the tanks remained unprotected. Nevertheless, the G4M2 became the major production model, with five other sub-types; the G4M2a Model 24 (bulged bomb doors, Kasei 25 engines); G4M2b Model 25 (Kasei 27s); G4M2c Model 26 (two 20mm, four 7.7mm guns); G4M2d (for flight-testing the Ne-20 turbojet engine); and G4M2e Model 24-J (four 20mm guns, one 7.7mm). Model 24-Js were later adapted as Ohka suicide aircraft carriers. Late in 1943, with continuing heavy losses, Mitsubishi built 60 G4M3a Model 34s and G4M3b Model 36s. These had a reduced fuel load of 968 Imp galls (4,4001) in fully protected tanks, in a much-redesigned wing. Total production of G4Ms (Allied code name Betty) was 2,479; many were converted to 20-seat troop transports towards the end of the war. The G4M3c Model 37 did not complete trials before VJ-day; the G4M4 was abandoned.

Ilyushin Il-2 Shturmovik

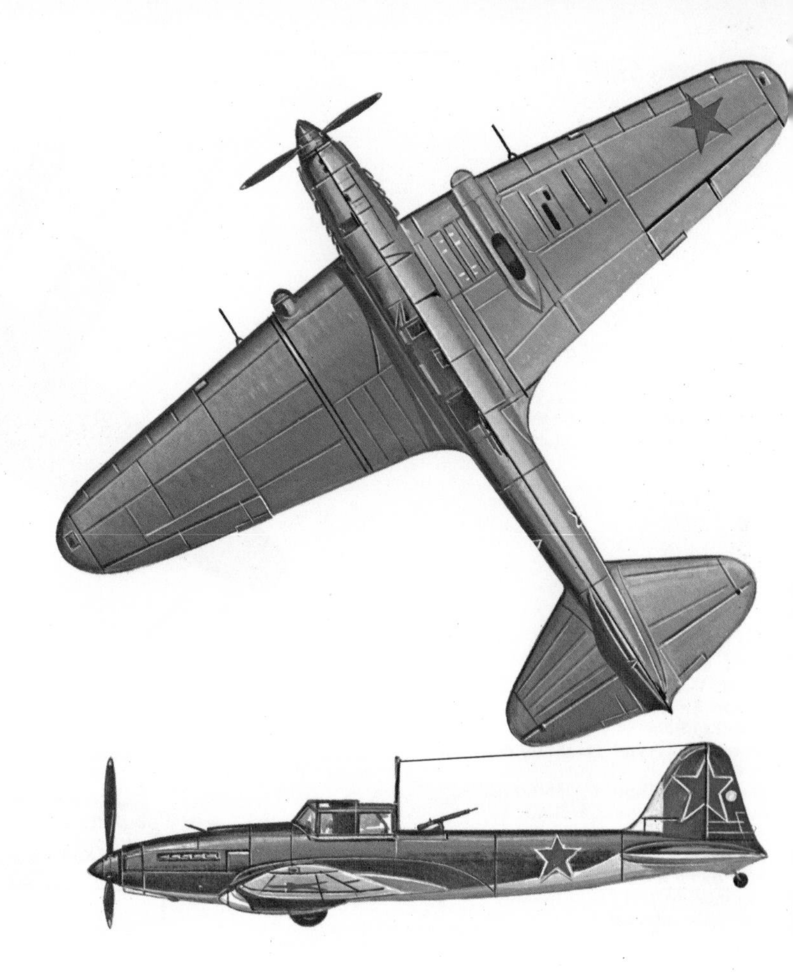

Ilyushin Il-2m3 *Shturmovik* of the Soviet Air Force, Eastern Front, 1944

Span: 47ft 10¾in (14.60m)
Length: 38ft 2⅝in (11.65m)
Weight, normal take-off: 12,147lb (5,510kg)
Engine: 1,770hp Mikulin AM-38F 12-cyl V-type
Max speed: 251mph (404kmh) at 4,920ft (1,500km)
Operational ceiling: 19,685ft (6,000m)
Range, normal: 373 miles (600km)
Armament: 2×23mm VY cannon; 2×7.62mm Shko machine-guns; 1×2.7mm UBT machine-gun
Max bomb load: 882lb (400kg) internally; plus 440lb (200kg) or 4×82mm RS-82 RPs externally

Probably the most advanced and most effective ground attack aircraft to serve during World War 2, the Il-2 Shturmovik, like the Ju 87 Stuka, introduced a new word into combat aircraft terminology. After studying proposals by a number of design teams, those of the Ilyushin bureau were accepted, materialising in the BSh-2 (or TsKB-55) prototype, which first flew on 30 December 1939. About 15 per cent of its total weight consisted of armour-plate protection for the engine, fuel and cooling systems and two-man crew. Tests with the first two prototypes indicated insufficient power and a lack of longitudinal stability. The original 1,370hp AM-35 engine was therefore replaced in a modified prototype (TsKB-57) by the new 1,680hp AM-38 which offered much greater power for take-off and low altitude flying. The TsKB-57 was a single-seater, with improved armament and capable of carrying several alternative external warloads; it first flew on 12 October 1940. This version entered production, as the Il-2, in the following spring, and carried out its first operational engagements in summer 1941. Output of aircraft and engines increased rapidly, the number of Il-2s in service being quadrupled by mid-1943. By this time a modified version was in production, in which a second crew member was restored to man a rear firing gun, and which had improved take-off performance, manoeuvrability and anti-tank weapons. This model, designated Il-2m3, entered production in mid-1942 and became operational in the following October. By early 1943 the 2-seat version was scoring heavily in air-to-air combat even against the Bf 109, and with masterly understatement the official trials report declared that it could 'be introduced with advantage into ground attack units'. The improved anti-tank weapons and 37mm cannon carried by later-production Il-2m3s maintained the aircraft's effectiveness even against the new German tanks in summer 1943; flying performance was sustained by introduction of the 1,750hp AM-48F engine in later production batches. The total quantity built – an estimated 35,000 – would alone make the Il-2 an outstanding aeroplane, but its intrinsic merits had no need of statistical support, and its achievements were their own recommendation.

Consolidated B-24 Liberator

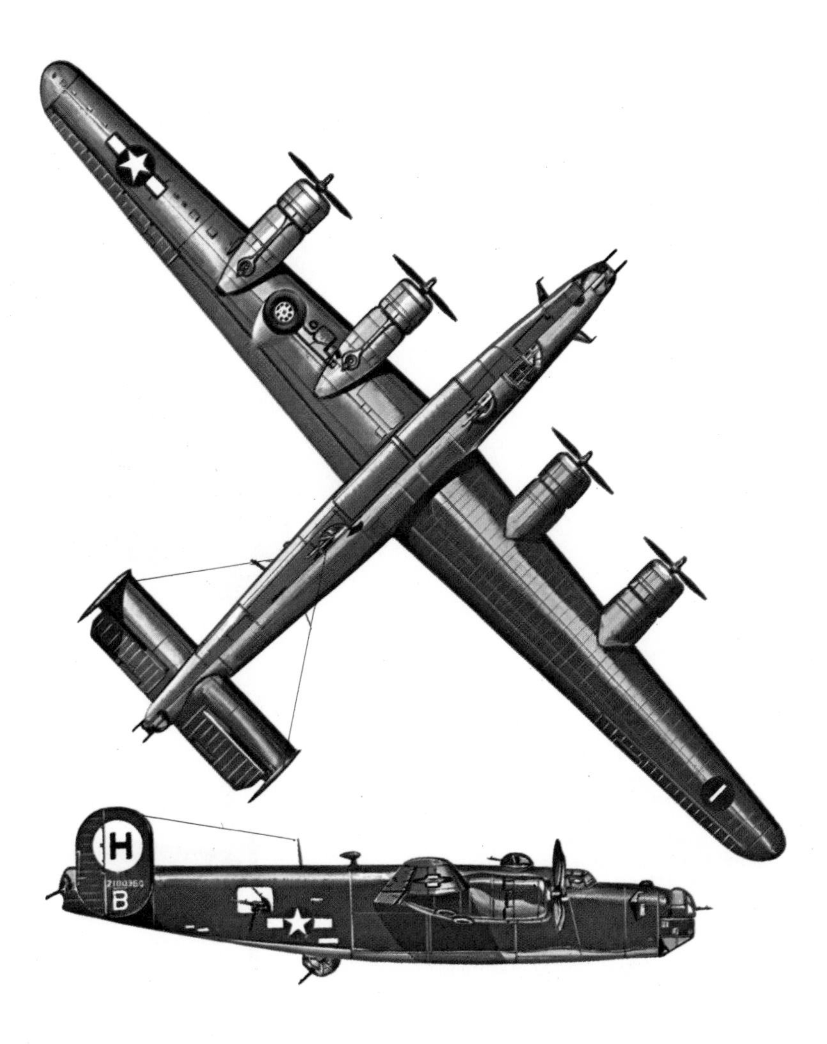

Convair B-24J-95-CO of the 448th BG, US Eighth Air Force, UK, *circa* November/ December 1943.

Span: 110ft 0in (33.53m)
Length: 67ft 2in (20.47m)
Weight: 56,000lb (25,401kg)
Engines: 4×1,200hp Pratt & Whitney R-1830-65 Twin Wasp 14-cyl radials
Max speed: 290mph (467kmh) at 25,000ft (7,620m)
Operational ceiling: 28,000ft (8,535m)
Range at max overload weight: 2,100 miles (3,380km)
Armament: 10×0.50in Browning machine-guns
Max bomb-load: 5,000lb (2,268kg) internally

The Liberator was built in greater numbers and more variants than any other World War 2 US aircraft and served in more combat theatres, over a longer period, than any heavy bomber. The exceptionally high aspect ratio Davis wing gave it prodigious range and the capacious fuselage permitted a large bomb or cargo-load. The XB-24 Consolidated Model 32 (prototype) flew on 29 December 1939, followed by seven service trials YB-24s and 36 production B-24As. The French ordered 120, which were diverted to Britain, the first few, designated LB-30 or -30A, being used by BOAC for transatlantic ferrying; 20 went to RAF Coastal Command as Liberator Mk Is, with ASV radar. B-24A deliveries to the USAAF began in June 1941, followed by nine B-24Cs with turbo-supercharged engines and revised armament. The first US version to serve in the bomber role was the B-24D, similar to the C except for R-1830-43s and increased gross weight; production, including ten by Douglas, totalled 2,738; the RAF received 260 as Mks III and IIIA and 122 as Mk Vs with radar and Leigh searchlights for Coastal Command. In mid-1943 the USAAF anti-submarine patrol B-24Ds were transferred to the USN and redesignated PB4Y-1. Convair (as Consolidated became), Douglas and Ford built 791 B-24Es, distinguishable from the D by different propellers: North American built 430 B-24Gs, some with a powered nose turret. An Emerson nose turret and R-1830-65s characterised the H; Convair, Douglas and Ford built 3,100, and then, with North American, manufactured 6,678 B-24Js, with Motor Products nose turret and Briggs ventral ball turret. The RAF received 1,278 as Mks VI (bomber) and VII (general reconnaissance); 977 went to the USN, designated PB4Y-1, most with radar replacing the ventral turret. Armament variations characterised the B-24L and M (1,667 and 2,593 respectively, by Convair and Ford). Ford completed seven single-finned YB-24Ns, and 46 similar RY-3s, for the USN, before contracts for over 5,000 B-24Ns were cancelled in May 1945. The Liberator shared the day bombing of Europe with the Boeing B-17 but was even more prominent in the Pacific where its range was particularly valuable. It gave considerable service as a transport, 276 B-24Ds being completed as C-87s; and some 100 B-24s became F-7 PR aircraft. The RCAF and Commonwealth air forces operated considerable numbers. Liberator production, ending on 31 May 1945, totalled 18,188.

Focke-Wulf Fw 190

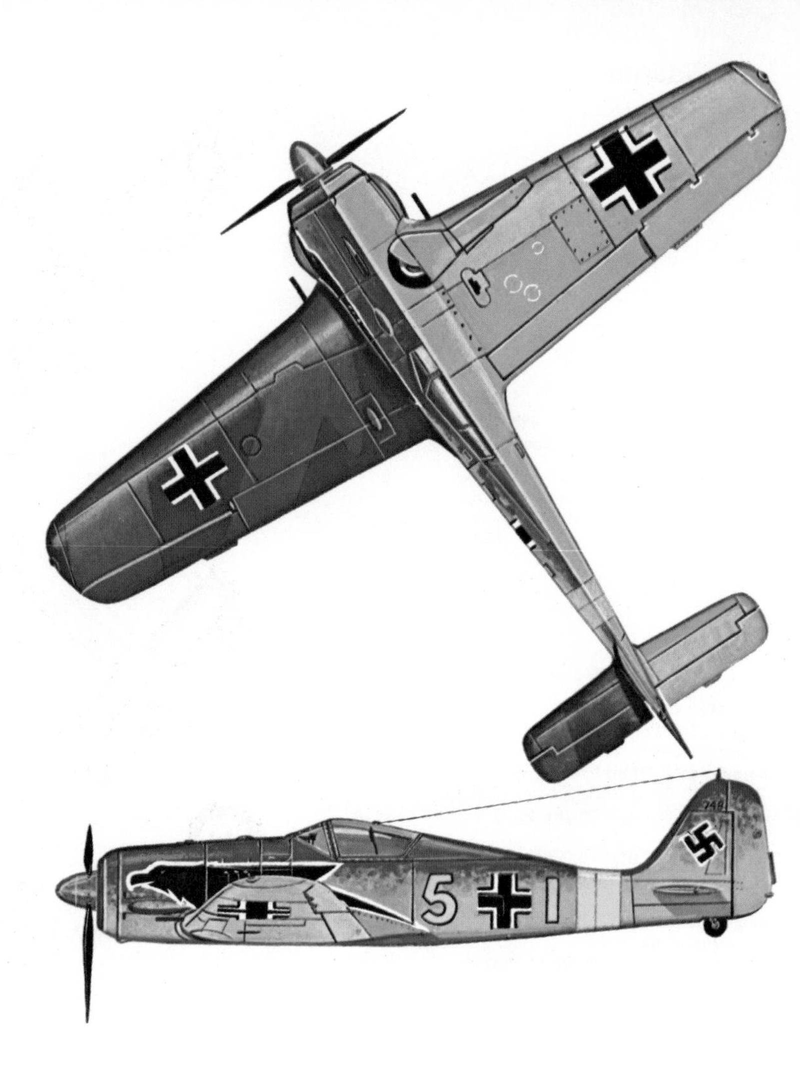

Focke-Wulf Fw 190A-4 of 9/*JG* 2 *Richthofen, Luftwaffe,* Vannes, France, February 1943

Span: 34ft 5⅜in (10.50m)
Length: 28ft 10½in (8.80m)
Weight, normal take-off: 8,378lb (3,800kg)
Engine: 1,700hp BMW 8010-2 14-cyl radial (2,100hp with power boost)
Max speed: 416mph (670kmh) at 20,590ft (6,275m)
Operational ceiling: 37,400ft (11,400m)
Range, normal: 497 miles (800km)
Armament: 2×7.9mm MG 17 machine-guns; 1×20mm MG151 cannon; 1×20mm MGFF cannon

The Fw 190, designed by Dipl-Ing Kurt Tank, was one of the technically most advanced, and operationally most eminent, fighter/fighter-bombers of 1939-45. One of two designs submitted by Focke-Wulf in response to a 1937 RLM specification, the prototype Fw 190V1 (D-OPZE) flew on 1 June 1939, powered by a 1,550hp BMW 139 radial, as was the second prototype. Subsequently, the larger 1,600hp BMW 801 was substituted. In 1940 18 pre-production Fw 190A-0s were ordered, most with a 3ft 3¼in (1.00m) increased in span that became standard. Delivery began in June 1941 with 100 A-1s. Armament was increased to six guns in the A-3, powered by a 1,700hp BMW 801Dg. The Fw 190 was used in low-altitude hit-and-run raids over southern England in 1941-42. By the end of 1942 nearly 2,000 had been built and the Fw 190 was serving in North Africa and on the Russian Front in even greater numbers than in Europe. Variants included the A-4; the A-4/U8 with fewer guns but carrying drop-tanks and a 500kg (1,102lb) bomb load, and the rocket-carrying A-4/R6; the A-5, for night fighting and close-support duties; the A-6 and A-7 had increased firepower; the A-8, A-9 and A-10 were mostly fighter-bombers, with different BMW 801 versions. A few Fw 190B and C prototypes were completed, with supercharged inverted-V DB603s. They were discarded in favour of the Fw 190D powered by the liquid-cooled 1,776hp Junkers Jumo 213A-1 whose annular radiator duct presented a radial-engined appearance and characterised by its longer nose and rear fuselage and (on the D-1) increased fin area. The initial D-0 and D-1 aircraft were evaluated in spring and summer 1943. The first major production D was the D-9, an interceptor, entering service in 1943; subsequent versions, equipped for ground attack, included the D-11, D-12 and D-13. Following the D was the Fw 190G fighter-bomber which could carry up to 1,800kg (3,968lb) of bombs. The F followed. Both were powered by the BMW 801D. Total Fw 190 production, excluding prototypes, was 20,051, over 6,500 of which were fighter-bomber variants. The DB603-engined Ta 152, developed from the Fw 190D, succeeded it in production, but served only in small numbers before the war ended.

Lockheed P-38 Lightning

Lockheed P-38J-15-LO Lightning of the 55th FS, 20th FG, US Eighth Air Force, UK, early 1944

Span: 52ft (15.85m)
Length: 37ft 10in (11.53m)
Weight, normal take-off: 17,500lb (7,938kg)
Engines: 2×1,425hp Allison V-1710-89/91 12-cyl V-type
Max speed: 414mph (666kmh) at 25,000ft (7,620m)
Operational ceiling: 44,000ft (13,410m)
Range (on internal fuel): 1,175 miles (1,880km)
Armament: 1×20mm Hispano M2 cannon; 4×0.50in Colt-Browning machine-guns
Max bomb-load: 3,200lb (1,452kg); or 10×5in rocket projectiles

Few would dispute the P-38's claim to epitomise the successful realisation of the long-range tactical fighter in World War 2. Design work began early in 1937 to meet an exacting USAAC requirement. Lockheed's Model 22 won the competition and in June 1937 one prototype, designated XP-38, was ordered. It flew on 27 January 1939, followed on 16 September 1940 by the first of 13 YP-38 evaluation aircraft with more powerful V-1710 engines and nose armament of four machine-guns and a 37mm cannon. Delivery of production P-38s began in June 1941; 30 were built, one modified to an XP-38A with a pressurised cabin. The next production model was the P-38D (36 built), with self-sealing fuel tanks, and airframe modifications. The RAF ordered the type in 1940, naming it Lightning, but only three Mk Is with non-supercharged engines were delivered; a contract for 524 Mk IIs was cancelled. The remaining 140 Mk Is were repossessed as the P-322 by the USAAF, which also acquired the Mk IIs; many were later converted to P-38Fs or Gs. Meanwhile, the USAAF's next version, the P-38E, entered production (210 built) with a 20mm replacing the 37mm cannon. Increased power was the major improvement in the F and G models, enabling the carriage of external weapons or fuel tanks for the first time. Production of 527 P-38Fs and 1,082 P-38Gs, deliveries beginning during 1942, heralded a marked expansion in the P-38's deployment in the major theatres of the war. The P-38H of 1943 had further increased power; 601 P-38Hs were delivered. The 2,970 P-38Js were similar, but introduced 'chin' air cooler intakes; increased internal fuel raised endurance (with drop-tanks) to 12 hours. Even greater numbers were built of the rocket-carrying P-38L, with 1,600hp V-1710-111/113 engines and a maximum speed of 414mph (666kmh). Lockheed manufactured 3,810 P-38Ls; 2,000 were ordered from Vultee, who completed 113 before the remainder were cancelled when the Pacific war ended. P-38s converted for other duties included a few P-38M night fighters from P-38Ls, a few TP-38L conversion trainers, and the undesignated 'Droop Snoot' and 'Pathfinder' (formerly J or L models). The most widely used single photo-reconnaissance aircraft of World War 2, nearly 1,400 E, F, G, H, J and L models were converted to F-4s or F-5s.

Mitsubishi Ki-46 (Dinah)

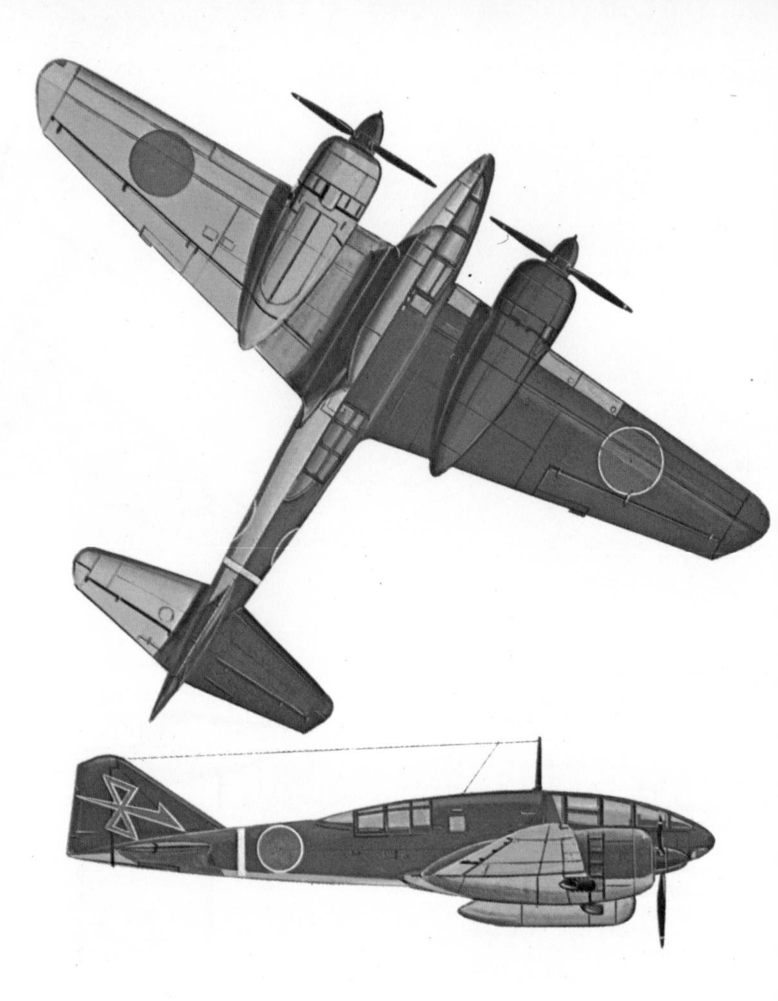

Mitsibishi Ki-46-III Model 3 of the 2nd Sqdn, 81st Group, JAAF, Burma 1944

Span: 48ft 2¾in (14.70m)
Length: 36ft 1in (11.00m)
Weight: 14,330lb (6,500kg)
Engines: 2×1,500hp Mitsubishi Ha-112-II 14-cyl radials
Max speed: 391mph (630kmh) at 15,685ft (16,000m)
Operational ceiling: 34,450ft (10,500m)
Range: 2,485 miles (4,000km)
Armament: None

The Ki-46, in terms of aerodynamics and performance one of the best Japanese aircraft to serve during World War 2, was designed by Tomio Kubo to a rigorous JAAF requirement issued in December 1937. The specification was for a fast, high altitude, long range, twin-engined reconnaissance aeroplane capable of speeds more than 50mph (80kmh) faster than the latest western single-engined fighters. These demands were met by ingenious weight-saving and excellent streamlining, and the first prototype, powered by 875hp Mitsubishi Ha-26-I engines, was flown in November 1939. The aircraft entered immediate production as the Ki-46-I, or Army Type 100 Model 1 command reconnaissance monoplane. The initial 34 aircraft were similarly powered, and were armed with a 7.7mm Type 89 machine-gun on a movable mounting in the rear cabin. They were used mostly for service trials and crew training, the main production version being the Ki-46-II Model 2, which first flew in March 1941; 1,093 Model 2s were built, 1,080hp Ha-102 engines enhancing performance. Delivery of these began in July 1941, initially to JAAF units in Manchuria and China. The Ki-46 (code named Dinah by the Allies) subsequently appeared in virtually all theatres of the Pacific war. A few Ki-46-IIs were also used by the IJN, and others were converted in 1943 to Ki-46-II-Kai operational trainers, with a second cockpit behind the pilot's cabin. In December 1942 two prototypes were flown of the Ki-46-III, and 609 Model 3s were completed. The Ki-46-IIIa featured a modified nose canopy, eliminating the step in front of the pilot's cabin, carried additional fuel and dispensed with the dorsal gun. Four prototypes were also built in 1945 of the Ki-46-IVa, basically similar but with turbo-supercharged Ha-112-IIru engines, but no production was undertaken. All the 1,742 Ki-46s were built by Mitsubishi at Nagoya and, later, at Toyama. In 1944, however, a substantial quantity of Model 3s began to be converted to Ki-46-III-Kai fighters by the Army Aeronautical Research Institute at Tachikawa for Japanese home defence. Conversion involved a 'stepped' nose, broadly similar to the original one, housing two 20mm Ho-5 cannon and, between the front and rear crew cabins, a 37mm Ho-203 cannon fixed to fire forward and upward. The Ki-46-IIIb, of which a few were built for ground attack, was similar, but omitted the dorsal cannon. Projected fighter variants included the Ki-46-IIIc (twin Ho-5s) and Ki-46-IVb (two nose-mounted Ho-5s).

Macchi C202 Folgore

Macchi C202 *Serie* XI Folgore of the *353° Squadriglia, 20° Gruppo, 51° Stormo Coccia Tenestre,* Regia Aeronautica, Monserrato, Italy, *circa* July 1943

Span: 34ft 8½in (10.58m)
Length: 29ft 0⅜in (8.85m)
Weight, normal take-off: 6,459lb (2,930kg)
Engine: 1,075hp Alfa Romeo RA 1000 RC 41 12-cyl invertyed V-type (licence-built DB601A-1)
Max speed: 370mph (595kmh) at 19,685ft (6,000m)
Operational ceiling: 37,730ft (11,500m)
Range, normal: 475 miles (765km)
Armament: 2×12.7mm Breda-SAFAT and 2×7.7mm Breda-SAFAT machine-guns

The first single-seat fighter designed for Aeronautica Macchi by Ing Mario Castoldi, the C200 Saetta (Lightning), was developed under the *Regia Aeronautica* re-equipment programme during the mid-1930s. Its contours were spoiled only by its bulky 850hp Fiat A74 RC38 radial. The first prototype flew on 24 December 1937. During 1938 an initial production order was placed for 99 with enclosed cockpits and more powerful A74s. Delivery began in October 1939. First appearing in World War 2 over Malta, the C200 subsequently served wherever Italian forces operated. Approximately 1,200 were built; most had open cockpits, preferred by Italian pilots. Although not fast or well-armed, it was well-built, extremely manoeuvrable and withstood considerable battle damage or climatic extremes. Attempts to improve the performance of the C200 began as early as 1938, when Macchi produced the C201 by redesigning the fuselage for a 1,000hp Fiat A76 RC40 radial. This engine was abandoned and the C201 was test-flown with an A74; in 1941 one C200 was refitted with a Piaggio P.XIX, but both were discarded in favour of the more promising C202. In 1940 a German DB601A-1 liquid-cooled inverted-V engine was installed in a C200 to create the prototype C202. It flew on 10 August 1940 and the aerodynamic and performance advances were such that immediate production was authorised. Initially, C202s had imported DB601s, but licence production began in Italy as the Alfa Romeo RA1000 RC41. C200 and C202 production continued in parallel, the first examples of the C202 Folgore (Thunderbolt) entering *Regia Aeronautica* service in summer 1941. At first, they carried similar armament to the Saetta, but later had two additional 7.7mm wing guns; one was tested with a 20mm Mauser MG151 cannon in a fairing beneath each wing. The Folgore is generally considered the most effective Italian fighter of the war. Serving in the Mediterranean, North Africa, and on the Eastern Front, it remained in production, though restricted by engine output, until the Italian armistice in September 1943. Macchi built 392, and Breda about 1,100. The C205V Veltro (Greyhound) was a much-improved development with a 1,475hp DB605A, but was too late to take a major part in the war.

Hawker Typhoon

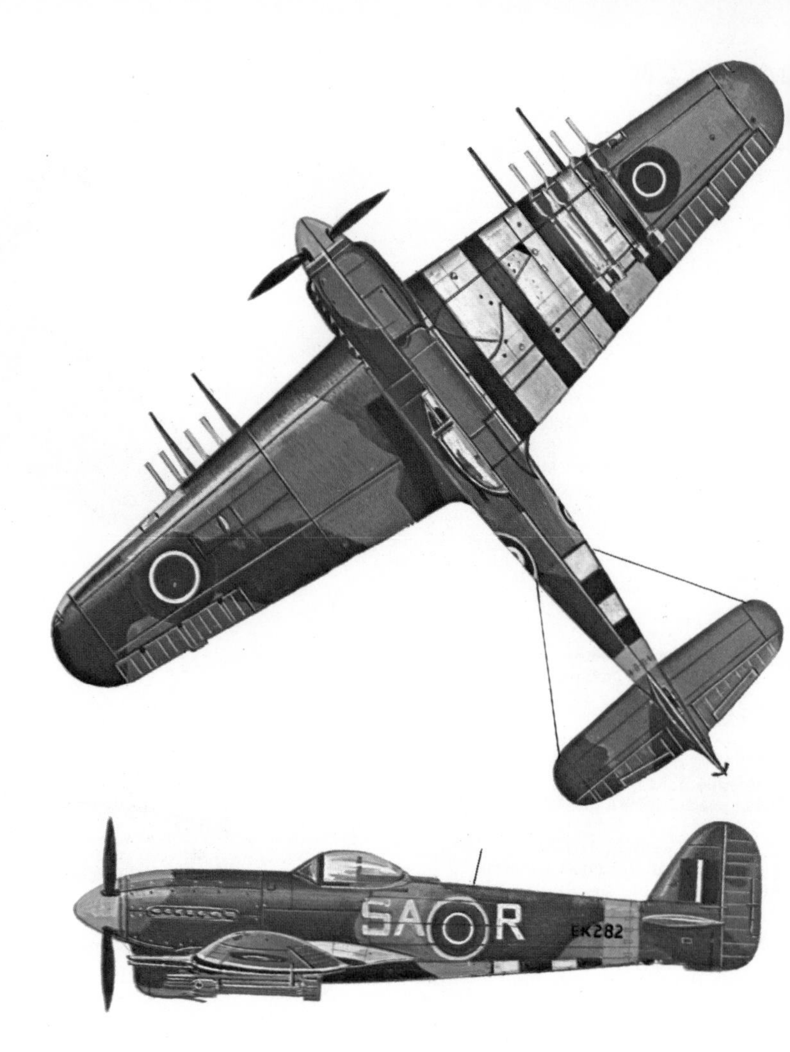

Hawker Typhoon Mk IB of No 486 Sqdn, RNZAF, UK, summer 1944

Span: 41ft 7in (12.67m)
Length: 31ft 10¾in (9.72m)
Weight: 12.905lb (5,853kg)
Engine: 2,200hp Napier Sabre IIB 24-cyl H-type
Max speed: 409mph (658kmh) at 10,000ft (3,050m)
Operational ceiling: 34,000ft (10,365m)
Range with drop-tanks: 910 miles (1,465km)
Armament: 4×20mm Hispano cannon
Max bomb load: 2,000lb (454kg); or 8× 60lbs (27kg) rocket projectiles RPs

The Typhoon was developed by Sydney Camm's design staff at Hawker Aircraft in response to Air Ministry Specification F18/37, which was for an interceptor capable of combating such heavily armed and armoured escort fighters as the Messerschmitt Bf 110. Such an aeroplane was inevitably heavier than either the Hurricane or the Spitfire, and to provide a comparable performance the powerplants selected were the new Napier Sabre H-type inline engine and the X-type Rolle-Royce Vulture, both of which promised to develop some 2,000hp. Prototypes were completed with both types of engine: the Vulture-engined design, named Tornado, was abandoned when Vulture production was curtailed. With the Sabre engine, the aircraft was named Typhoon, and the first of two prototypes (P5212) flew on 24 February 1940. Typhoons entered service in September 1941. Early service trials and squadron experience were far short of being satisfactory, and the Typhoon's future career might soon have ended but for the appearance in 1941 of the Focke-Wulf Fw 190 intruder raids across the English Channel. The Fw 190 could outmanoeuvre all other British fighters, including the Spitfire Mk V, and the Typhoon was the only effective means of stopping it. Early Typhoon Mk IAs carried six 0.303in Browning guns in each wing, but the Mk IB had four wing-mounted 20mm cannon, which became the Typhoon's regular fixed armament. The fighter was unspectacular at altitude, but its clashes with the Fw 190 had revealed outstanding strength and agility at low level, and from this stemmed the type's widespread use and success as a ground attack aircraft. After extensive weapons trials during 1942, Typhoons were fitted for operational use in 1943 with underwing rails for eight rocket projectiles, the chief weapon employed by the type. Before and after the invasion of Europe, rocket-armed Typhoons attacked land and sea targets in the Channel and in Belgium, France and the Netherlands. A total of 3,330 Typhoons was built, all by Gloster except for the two prototypes, five Mk IAs and ten Mk IBs. The Mk IB was the major version, over 3,000 being completed with Sabre IIA, IIB or IIC engines, some 60 per cent of this total having bubble-type canopies replacing the original frame-type hood and cockpit door.

De Havilland Mosquito

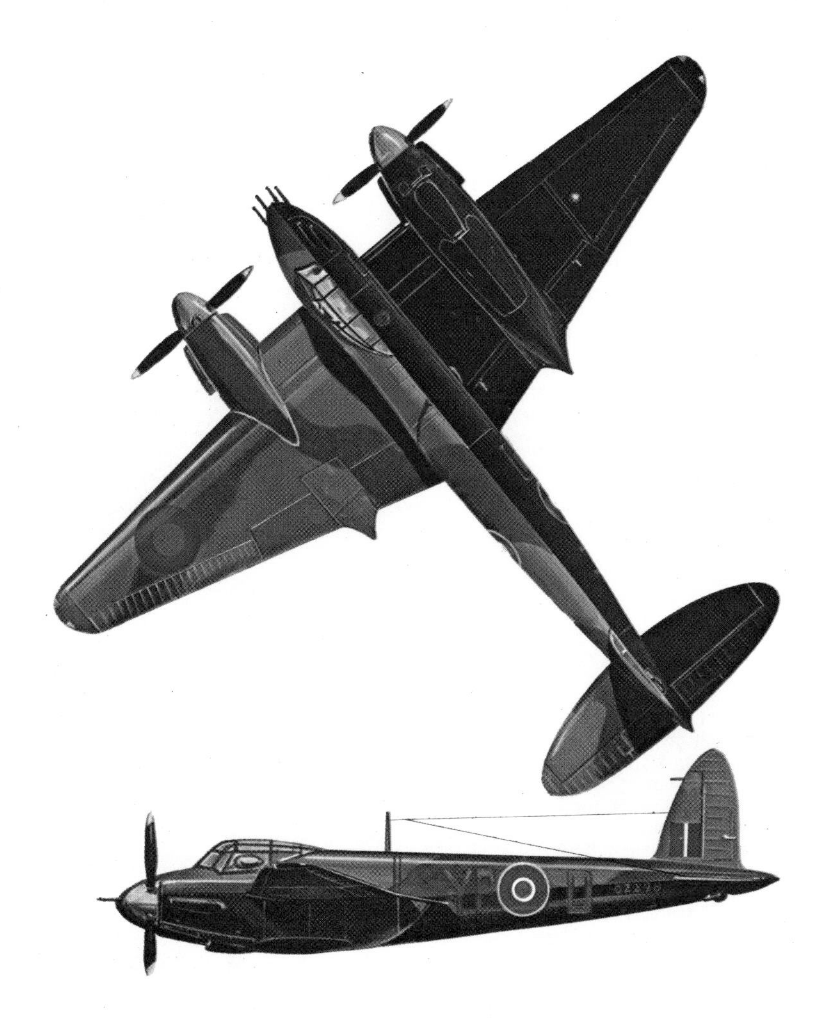

De Havilland Mosquito NF Mk II of No 23 Sqdn, RAF, Malta, January 1943

Span: 54ft 2in (16.51m)
Length: 40ft 4in (12.29m)
Weight, normal take-off: 18,100lb (8,210kg)
Engines: 2×1,460hp Rolls-Royce Merlin 21/23 12-cyl V-type
Max speed: 354mph (570kmh) at 14,000ft (4,267m)
Operational ceiling: 34,500ft (10,515m)
Range normal: 1,520 miles (2,446km)
Armament (NF Mk II): 4×20mm Hispano cannon; 4×0.303in Browning machine-guns
Max bomb load: None in NF MkII; max for B Mk XVI 4,000lb (1,814kg)

The DH98 Mosquito was conceived in 1938 as an unarmed day bomber fast enough to outrun enemy fighters. Not until March 1940 was official interest firm. The first and second prototypes were to bomber and PR configuration, flying on 25 November 1940 and 10 June 1941 respectively. The initial 50 included ten PR Mk I and ten B Mk IV; the first operational sortie was flown by a PR Mk I on 20 September 1941. B Mk IVs entered service in May 1942; 273 Mk IVs were built. During the war, the three UK de Haviland factories built 4,444; Airspeed 122; Percival 245; Standard Motors 165; de Havilland Australia 208 (Mks 40-43), and de Havilland Canada 1,134 (Mks VII, 20-22 and 24-27), production totalling 6,710; 1,071 were completed after VJ-day. Mosquitos quickly established a reputation for excellent flying qualities, unequalled pin-point bombing and easily the lowest loss rate of any Bomber Command aircraft. Mk IVs originally equipped the Pathfinder Force, which later employed the high-altitude Mk IX and the Mk XVI, second most numerous variant (about 1,200 built); counterparts of these three marks were the principal wartime PR Mosquitos. From early 1944 the most popular weapon was the 4,000lb 'block-buster' bomb, carried in a bulged bomb bay retrospectively fitted to Mks IX and XVI and several Mk IVs. The third prototype, flying on 15 May 1941, was a night fighter with AI Mk IV radar in a 'solid' nose; 466 similar Mk IIs were built, deliveries beginning in January 1942. Later, 97 were converted to NF Mk XIIs with centrimetric AI Mk VIII radar; 270 NF Mk XIIIs, the Mk XII's production counterpart, followed. On radar-carrying night fighters nose machine-guns were omitted. Other specialist night fighters included the Mks XV, XVII (100 Mk II conversions), and XIX (220 built), the last two having US-manufactured AI Mk X. The most numerous version was the Mk VI, of which 2,718 were built during and after the war. Carrying standard fighter armament, the Mk VI could carry two 250lb or 500lb bombs, with two bombs or drop-tanks under wing. It entered service in spring 1943. Anti-shipping strike Mk VIs with eight 60lb RPs entered Coastal Command service early in 1944, preceded by 27 FB Mk XVIIIs, converted FB Mk VIs mounting a 57mm cannon in the nose. Principal Canadian and Australian FB Mk VI counterparts were the FB Mk 26 and FB Mk 40 with Packard-Merlins. The Mosquito served in all war theatres, including the Pacific from early 1944. Most of the 348 T Mk III trainers were delivered to the RAF, a few to the FAA and some were sold abroad.

Nakajima Ki-43 Hayabusa (Oscar)

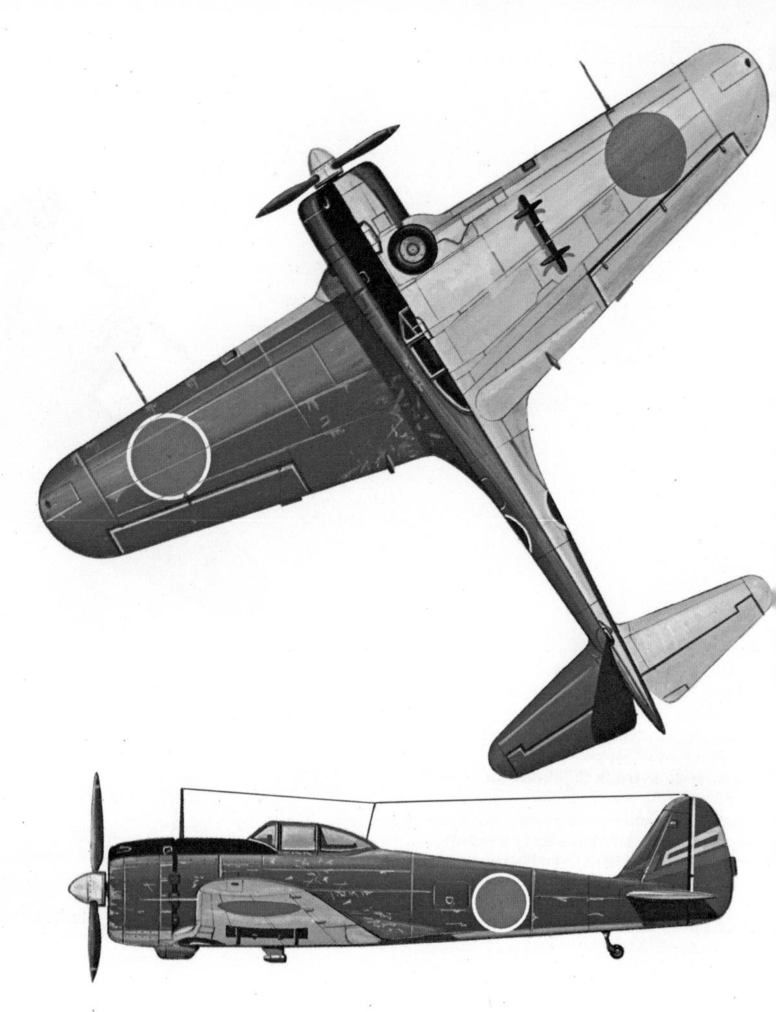

Nakajima Ki-43-IIIa Model 3A Hayabusa of the 20th Group, JAAF, home defence of Japan, 1944-45

Span: 35ft 6¾in (10.84m)
Length: 29ft 3⅛in (8.92m)
Weight: 6,746lb (3,060kg)
Engine: 1,190hp Mitsubishi Ha-112 14-cyl radial
Max speed: 342mph (550kmh) at 19,195ft (5,850m)
Operational ceiling: 37,400ft (11,400m)
Range: 1,988 miles (3,200km)
Armament: 2×12.7mm Type 1 machine-guns
Max bomb load: 440lb (200kg)

The Ki-43 was designed by Dr Hideo Itokawa in 1938 as a potential replacement for the JAAF's Ki-27 fighter. The first of three prototypes was flown early in January 1939 and had excellent speed and range, but was heavy on the controls. The first of the ten pre-production aircraft, appearing in November 1939, was a lighter aeroplane with increased wing area and 'combat flaps' that vastly improved its handling characteristics. It was quickly ordered in quantity and remained in production throughout the Pacific war; 5,751 were delivered. The initial version, the Ki-43-Ia Model 1A, entered service in October 1941. Fire-power was improved in later versions, the major early version being the Ki-43-Ic Model 1C with two fuselage-mounted 12.7mm guns. At the time of Pearl Harbor about 40 Ki-43s were in JAAF service and, although extremely popular as flying machines, combat experience soon revealed a need for greater armour protection and increased engine power. These appeared in the Ki-43-IIa Model 2A, built by Tachikawa factories in 1942-43, which was powered by the 1,105hp Sakae Ha-115 engine and could carry two 551lb (250kg) underwing bombs. This was succeeded in November 1943 by the Ki-43-IIb Model 2B, a clipped-wing variant of the Model 2A with greater manoeuvrability. Joint production by Nakajima and Tachikawa from December 1944 yielded the Ki-43-IIIa Model 3A (1,250hp Kasei Ha-112 engine), the last production version. Two prototypes were completed of the Tachikawa-developed Model 3B, with two 20mm cannon, but no production was achieved before VJ-day. The Hayabusa was encountered in particularly strong numbers during the battle for Leyte Island, and in the defence of the Kurile Islands north of Japan, but it served widely throughout all the mainland and island battle areas of south-east Asia, in suicide attacks during 1944-45, and in the final defence of the Japanese homeland. The Hayabusa (Pacific code name Oscar) was an excellent and versatile fighter, its only serious drawback being its lack of adequate armament.

Avro Lancaster

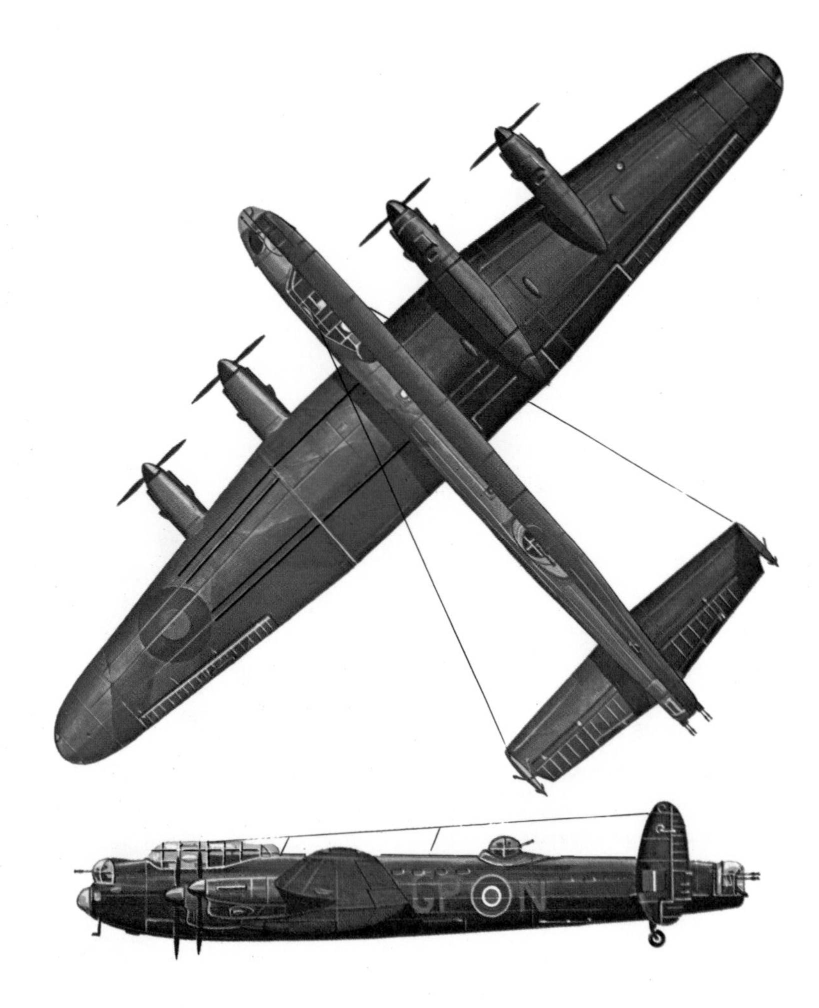

Avro Lancaster B Mk I of No 1661 Conversion Unit, RAF, February 1944

Span: 102ft 0in (31.09m)
Length: 69ft 4in (21.13m)
Weight: 68,000lb (30,844kg)
Engines: 4×1,280hp Rolls-Royce Merlin XX or 22 12-cyl V-type
Max speed: 287mph (462kmh) at 11,500ft (3,505m)
Operational ceiling: 24,500ft (7,465m)
Range with 12,000lb (5,443kg) bomb load: 1,730 miles (2,784km)
Armament: 8×0.303in Browning machine-guns
Max bomb load: 18,000lb (8,165kg) internally

The Lancaster became, in the admittedly partisan view of Air Chief Marshal Sir Arthur 'Bomber' Harris, the greatest single factor in winning World War 2 in Europe. It was a direct development of Avro's unsuccessful Manchester twin-engined bomber, its prototype (BT308) being originally named the Manchester Mk III, complete with triple tail unit, but having four 1,130hp Merlin X engines in place of the two Vulture engines of the Manchester Mk I and II. This prototype flew on 9 January 1941, followed just over nine months later by the first production Mk I, of which delivery began (to No 44 Sqdn) shortly after Christmas 1941. The first Lancaster combat mission came on 2 March 1942, laying mines in the Heligoland Bight and first bombing raid, on Essen, eight days later. The Lancaster Mk I, fitted successively with Merlin XX, 22 or 24 engines, remained the only version in service throughout 1942 and early 1943, and an eventual total of 3,440 Mk Is were completed by Avro, Armstrong Whitworth, Austin Motors, Metropolitan-Vickers and Vickers-Armstrongs. Avro completed two prototypes for the Lancaster Mk II powered by 1,725hp Bristol Hercules radial engines as a safeguard against possible shortages of Merlins. No such shortage arose, but 300 production Lancaster Mk IIs were built by Armstrong Whitworth. The other principal version, the Lancaster Mk III, was powered by Packard-built Merlin 28, 38 or 224 engines; apart from a modified bomb-aimer's window, this exhibited few other differences from the Mk I. Most of the 3,020 Mk IIIs completed were built by Avro, but 110 were manufactured by Armstrong Whitworth and 136 by Metropolitan-Vickers. Marks IV and V were extensively redesigned models that eventually became the Lincoln, while the small batch of Mk VIs were Mk IIIs converted to Merlin 85 or 87 engines in redesigned cowlings. The final British variant was the Mk VII, 180 being built by Austin with Martin dorsal turrets mounting twin 0.50in guns. In Canada, Victory Aircraft Ltd manufactured 430 Lancaster Mk Xs, which had Packard-Merlin 28s and were essentially similar to the Mk III. The Lancaster's bomb-carrying feats were legion. It was designed originally to carry bombs of 4,000lb (1,814kg), but successive modifications to the bomb bay produced the Mk I (Special) capable of carrying first 8,000lb (3,629kg) and then 12,000lb (5,443kg) weapons and, eventually, the 22,000lb (9,979kg) 'Grand Slam' armour-piercing weapon designed by Barnes Wallis. This remarkable engineer also designed the skipping bomb carried by the Lancasters of No 617 Squadron in their epic raid on the Möhne and Eder dams on 17/18 May 1943. By the end of the war Lancasters equipped no fewer than 56 Bomber Command squadrons.

Douglas C-47 Skytrain

Douglas C-47A Skytrain of *Eskadrille* 721, Royal Danish Air Force, *circa* 1965

Span: 95ft 6in (29.11m)
Length: 63ft 9in (19.43m)
Weight: 26,000lb (11,795kg)
Engines: 2×1,200hp Pratt & Whitney R-1830-92 Twin Wasp 14-cyl radials
Max speed: 229mph (368kmh) at 7,500ft (2,285m)
Operational ceiling: 24,000ft (7,315m)
Range, normal: 1,500 miles (2,415km)
Max pay load: 7,500lb (3,402kg) cargo; or 28 troops

First Douglas Commercial transports acquired by the US services were DC-2s (Army C-32A and C-34, Navy R2D-1), followed by 35 C-39s with DC-2 fuselages and DC-3 tail surfaces and outer wing panels. The principal wartime versions were the Twin Wasp-engined C-47 Skytrain, C-53 Skytrooper and USN R4D series, differing in minor detail and function. The first 953, delivered from December 1941, were C-47 troop or cargo transports; they were followed by 4,991 C-47As, and from 1943 by 3,108 C-47Bs, all by Douglas; 133 TC-47Bs were built for training. The Skytrooper, a troop transport, preceded the C-47 into service (October 1941); Douglas produced 193 C-53s, eight C-53Bs, 17 C-53Cs and 159 C-53Ds from 1941 to 1943. In 1945 17 C-117A VIP transports were delivered. Commercial airline DSTs (Douglas Sleeper Transports) or DC-3s impressed for war service included 36 designated C-48 to C-48C, 138 C-49 to C-49K, 14 C-50 to C-50D, one C-51, six C-52 to C-52C, two C-68 and four C-84. Over 1,200 C-47s were supplied under Lend-Lease to the RAF, as Dakota Mks I to IV. They entered service in Burma in June 1942. Roles included casualty evacuation and glider towing. Wartime production of military DC-3s, ending in August 1945, totalled 10,047, mostly built by Douglas. In addition to about 700 supplied to the USSR under Lend-Lease, some 2,000, as the Lisunov Li-2 (formerly PS-84), were built in the USSR. Licence production was also undertaken in Japan, Showa building 380 and Nakajima 70 L2D2s and L2D3s for the JNAF (code-named Tabby by the Allies). The number still operating with small airlines and over 60 air arms runs into thousands. Perhaps a greater tribute is that about two dozen variants were employed by the USAF in the 1970s. Some 20 were used in Vietnam as AC-47s for ground strafing, with three rapid-fire GE 7.62mm Miniguns in the port doorway. In wartime the C-47 has successfully carried as many as 74 men. The 'Super DC-3', with redesigned, square-cut wing and tail surfaces, uprated engines and other improvements, was built for the USAAF (17 C-117s); 98 earlier R4Ds were converted to R4D-8 'Supers' for the USN. Many Li-2s serve with pro-Soviet states.

Grumman TBF/TBM Avenger

Eastern-built TBM-3 Avenger of Air Group 38, US Navy, August 1945

Span: 54ft 2in (16.51m)
Length: 40ft 0⅛in (12.19m)
Weight: 18,250lb (8,278kg)
Engine: 1,900hp Wright R-2600-20 Cyclone 14-cyl radial
Max speed: 267mph (430kmh) at 15,000ft (4,570m)
Operational ceiling: 23,400ft (7,130m)
Range: 2,530 miles (4,072km)
Armament: 3×0.50in machine-guns; 2×0.30in machine-gun
Max bomb load: 1×22in torpedo or 1×2,000lb bomb internally

The Avenger was a pre-war design, two XTBF-1 prototypes of which were ordered by the USN in April 1940. The first made its maiden flight on 1 August 1941, by which time a substantial first order had been placed. The first production TBF-1s were delivered to VT-8 late in January 1942, and the Avenger made its combat debut early in the following June at the Battle of Midway. The aircraft had typical Grumman lines, the most noticeable feature being the very deep fuselage, which enabled the torpedo or bomb load to be totally enclosed. The TBF-1C had two wing-mounted 0.50in machine-guns in addition to the nose, dorsal and ventral guns of the original TBF-1, and could carry auxiliary drop-tanks. Both models were 3-seaters and were powered by the 1,700hp R-2600-8 engine. Up to December 1943 Grumman built 2,293 TBF-1/-1C Avengers, including the two original prototypes, one XTBF-2 and one XTBF-3; 402 of them were supplied to the Royal Navy as Avenger Mk Is (TBF-1B) and 63 to the RNZAF. The British aircraft were briefly known as Tarpon, but the US name was later standardised. Meanwhile, in the USA production had also begun in September 1942 by the Eastern Aircraft Division of General Motors, which built 2,882 as the TBM-1 and -1C, of which 334 went to the FAA as Avenger Mk IIs. The 'dash 2' variant was not built by either company, but Eastern completed a prototype and 4,664 TBM-3s with uprated Cyclone engines and their wings strengthened to carry rocket projectiles or a radar pod; 222 of these became the British Avenger Mk III. Further strengthening of the airframe produced the XTBM-4, but production of this model was cancelled when the war ended. This did not, however, end the Avenger's long and productive career: those of the USN were not finally retired until 1954, and postwar variants served with some foreign naval air forces for several years after this. During the major part of World War 2 the Avenger was the standard USN torpedo bomber, operating from carriers and shore bases, mostly in the Pacific theatre.

Republic P-47/F-47 Thunderbolt

Republic P-47D-21-RE Thunderbolt of the 62nd FS, 56 FG, USAAF, UK, May 1944

Span: 40ft 9⅜in (12.43m)
Length: 36ft 1¾in (11.02m)
Weight, normal take-off: 13,500lb (6,123kg)
Engine: 2,300hp Pratt & Whitney R-2800-21 Double Wasp 18-cyl radial
Max speed: 433mph (697kmh) at 30,000ft (9,144m)
Operational ceiling: 40,000ft (12,190m)
Range, normal: 640 miles (1,030km)
Armament: 8×0.50in Browning M-2 machine-guns
Max bomb load: 2,500lb (1,134kg)

In the light of the early air fighting in Europe Alexander Kartveli almost completely redesigned the XP-47 light fighter projected early in 1939, resulting in the XP-47B. It was almost twice as heavy and a Double Wasp radial engine replaced the XP-47's Allison in-line. In September 1940, 171 P-47Bs and 602 P-47Cs were ordered. The XP-47B flew on 6 May 1941. The B and C models were similar, but the C had a slightly longer fuselage to improve manoeuvrability. Thunderbolts entered USAAF service in March 1942, becoming operational with 8th Air Force units over Europe in April 1943 and in the Pacific theatre some two months later. Huge orders had been placed for the P-47D, which was initially a refined C. To this configuration, Republic manufactured 5,423 P-47Ds and Curtiss 354 designated P-47G. From the P-47D-25 the cockpit view was vastly improved by cutting down the rear fuselage and fitting a 'teardrop' canopy. The weight saved allowed extra fuel to be carried. Production batches from P-47D-27 had a dorsal fin fillet to offset the reduced keel area. Bubble-canopied P-47Ds served widely as a fighter and fighter-bomber in Europe; 8,179 were completed at Farmingdale and Evansville. The RAF received 240 Thunderbolt Mk Is (early P-47D) and 590 Mk IIs (later P-47D); 203 were allocated to the Soviet Air Force under Lend-Lease and 88 to Brazil. The next production model (following various experimental variants) was the P-47M; this utilised the 2,800hp R-2800-57 with which the XP-47J had flown at 504mph (811kmh), in the P-47D airframe. An improvised version produced hastily to counter the V1 flying-bomb attacks on Britain, only 130 were built. The last and heaviest production Thunderbolt was the P-47N, a very long-range escort and fighter-bomber variant; Republic built 1,816. Production, ending in December 1945, totalled 15,660 aircraft. About two-thirds survived the war. The Thunderbolt was one of the principal types supplied under the 1947 Rio Pact to many Latin American countries, including Bolivia, Brazil, Colombia, the Dominican Republic, Ecuador, Honduras, Nicaragua and Peru, a few serving until the late 1960s.

North American P-51/F-51 Mustang

North American F-51D of the Guatemalan Air Force

Span: 37ft 0¼in (11.28m)
Length: 32ft 3in (9.83m)
Weight: 9,800lb (4,445kg)
Engine: 1,695hp Packard Merlin V-1650-7 12-cyl V-type
Max speed: 439mph (706.5kmh) at 25,000ft (7,620m)
Operational ceiling: 41,900ft (12,770m)
Range: 950 miles (1,529km)
Armament: 6×0.5in machine-guns
Max bomb load: 10×HVAR rockets; or 6×HVAR rockets and 2×500lb or 2×1,000lb bombs externally

The Mustang was conceived to meet a British requirement for a high-speed fighter posed in April 1940, and was developed by a design team led by Raymond Rice and Edgar Schmued. With manufacturer's designation NA-73, the prototype flew on 26 October 1940 powered by a 1,100hp Allison V-1710-F3R. The initial British orders were for 620 Mustang Mk Is, the first reaching the UK in November 1941. Two were evaluated by the USAAF as XP-51s and ordered 150 for Lend-Lease to the RAF as Mustang Mk IAs but 55 were repossessed by the USAAF and converted to F-6A photo-reconnaissance aircraft; two others became XP-78s (later XP-51Bs) when fitted in 1942 with Packard-built Merlins. This followed British experiments with Merlin 60 series engines fitted in five Mustang Mk Is. The Merlin became the standard powerplant, but before this the USAAF received 500 Allison-engined A-36As, a ground attack variant and 310 P-51As. The RAF received 50 P-51As (Mustang Mk II), and 35 others were converted to F-6Bs. The A-36A was briefly named Invader and the P-51 Apache, but the British name Mustang was adopted for all variants. First Merlin-engined models were the P-51B and P-51C (Mustang Mk III), US production of which totalled 3,738; 910 were supplied to the RAF fitted with bulged canopies to improve visibility. Conversions of P-51B/Cs into F-6Cs totalled 91. The P-51D introduced a cut-down rear fuselage and a 'teardrop' canopy. Production totalled 9,293 of this model and similar P-51K; 876 became RAF Mk IVs and 299 became reconnaissance F-6Ds or F-6Ks. Last production model was the P-51H; 555 were completed in 1945. Contracts for over 3,000 Mustangs were cancelled at the war's end. The first RAF Mustangs became operational as armed tactical reconnaissance aircraft in May 1942, while from December 1943 P-51Bs flew as escorts to US 8th Air Force bombers over Europe. The Mustang figured largely in the Allied campaigns in North Africa, against V1 flying bombs over Britain in 1944 and as escort during the B-29 bombing raids of 1944-45 against Japan. Unquestionably, it was one of the greatest and most versatile fighters ever built, and a firm favourite with all who flew it. When Mustangs were phased out of US service hundreds were sold to Latin American signatories to the 1947 Rio Pact, some for a nominal one dollar. Between 50 and 70 F-51Ds remained in service with Bolivia, the Dominican Republic, Guatemala, Haiti, Honduras, Indonesia and Nicaragua in the mid-1960s.

Kawasaki Ki-61 Hien (Tony)

Kawasaki Ki-61-Ib Model 1B Hien of the 68th *Sentai* (Fighter Group), JAAF, New Britain, *circa* spring 1944.

Span: 39ft 4½in (12.00m)
Length: 28ft 8½in (8.75m)
Weight, normal take-off: 6,504lb (2,950kg)
Engine: 1,175hp Kawasaki H-40 12-cyl inverted V-type
Max speed: 368mph (592kmh) at 15,945ft (4,860m)
Operational ceiling: 38,060ft (11,600m)
Range, normal: 373 miles (600km)
Armament: 4×12.7mm Type 1 machine-guns

Two designs were formulated by Dr Takeo Doi to meet a February 1940 JAAF fighter requirement: the Ki-60 'heavy' fighter and the lightweight Ki-61. Three Ki-60 prototypes were completed, but it was discarded for the more conventional Ki-61. Twelve Ki-61 prototypes were built, all essentially similar. The first flew in December 1941 powered by a 1,100hp Ha-40 liquid-cooled engine, evolved in Japan from the German DB601A; this later led to the erroneous assumption that the Ki-61 was a licence-built development of the Messerschmitt Bf 109. The initial production model, delivery of which began in August 1942, was the Ki-61-I Model 1, which was armed with two 12.7mm and two 7.7mm guns. Successive improvements in armament led to the Model 1A (two 7.7mm fuselage guns; two 20mm wing-mounted Mauser cannon), Model 1B (four 12.7mm guns), Model 1C (two fuselage 12.7mm; two 20mm Ho-5 in the wings) and Model 1D (two fuselage 12.7mm; two 30mm in the wings). Total production of the Ki-61-I series fighters was 2,734. Following its first operational appearance in New Guinea in April 1943, the Hien (Allied code name Tony) was encountered in virtually all battle areas of the Pacific war. It was particularly prominent around Rabaul, in the battle for Leyte Island, and in the home defence of Japan. In September 1942, to offset certain maintenance difficulties encountered with the Ha-40 engine, Kawasaki began to evolve the Ki-61-II, utilising the new Ha-140 which promised to develop 1,450hp. The first Ki-61-II Model 2, completed in August 1943, featured a lengthened fuselage, modified canopy and a ten per cent increase in wing area. However, difficulties with the Ha-140 engine and associated structural problems prevented more than another seven being completed by January 1944. Attention was then devoted to the Ki-61-IIa Model 2A, with strengthened airframe and the wings and armament of the Ki-61-Ic. The Model 2B was similar but had four 20mm guns. However, after only 31 Model 2As and 2Bs had been completed, output was slowed down by the lack of powerplants. Another 374 Ki-61 airframes were completed, but only 99 received their intended powerplants and more than one-third of those were destroyed in air attacks before delivery. The remainder were eventually fitted with 1,500hp Mitsubishi Ha-112-II radial engines to become Ki-100s, and in this form were so successful that further development of the proposed Ki-61-III became unnecessary.

Vought F4U Corsair

Vought F4U-1A Corsair of VF-17, US Navy, summer 1943

Span: 40ft 11¾in (12.49)m
Length: 33ft 4½in (10.17m)
Weight, normal take-off: 11,093ft (5,032kg)
Engine: 2,000hp Pratt & Whitney R-2800-8 Double Wasp 18-cyl radial
Max speed: 417mph (671kmh) at 19,900ft (6,065m)
Operational ceiling: 36,900ft (11,245m)
Range, normal: 1,015 miles (1,633km)
Armament: 6×0.50in Browning machine-guns
Max bomb load: 1,000lb (227kg)

Originating with the Vought-Sikorsky Division of United Aircraft Corpn, the Vought V-166B, designated XF4U-1, flew on 29 May 1940. It was the first US warplane to exceed 400mph. It remained in production until 1953; 12,571 were built. The initial USN contract was for 584 F4U-1s; delivery began in September 1942 to USMC/USN land-based squadrons, due to difficulties in operating from carriers, and the first operational missions were flown by VMF-124, USMC in February 1943. The gull-wing was used to avoid the need for long undercarriage legs to clear the large-diameter propeller. But the far-aft cockpit position gave a poor view when landing; hence, from the F4U-1, a raised canopy was introduced. The Vought F4U-1C had four 20mm wing cannon instead of six machine-guns; the F4U-1D and the similar Goodyear FG-1D had a water-injection R-2800-8W and provision for eight underwing RPs or two 1,000lb (454kg) bombs. Brewster, after manufacturing 735 F3A-1s, ceased production in 1944. Goodyear built 4,014 FG-1s and -1Ds and Vought 4,669 F4U-1s to -1Ds. The FAA received 1,977 as Corsair Mks I to IV, the RNZAF 425. Mks II to IV had each wing clipped by 8in (20.3cm) for stowage aboard carriers, and preceded US F4Us into carrier service, entering action in April 1944. In 1943, 12 F4U-1s were modified to F4U-2s with four wing guns and radar in a starboard wingtip fairing; others became F4U-1P PR aircraft. The F4U-4 (Goodyear FG-4) had six 0.50in wing guns and a 2,100hp R-2800-1W. Delivery began late in 1944. Despite large cuts in orders after VJ-day, Vought completed 2,356 F4U-4s and Goodyear 200 FG-4s, including radar-equipped F4U-4E and -4N night fighters. Goodyear built five F2G-1s and five F2G-2s, with 3,000hp R-4360-4 Wasp Majors. During World War 2 US Corsairs operated mostly from land bases in the Pacific. The note of the airstream through the cooler inlets, and their 11:1 'kill ratio', led the Japanese to nickname them 'Whistling Death'. The postwar versions – F4U-5, AU-1 (F4U-6) and F4U-7 – served with distinction in the Korean War. Postwar, France's *Aéronavale* operated F4U-7s until 1964. Until the late 1960s F4U-5s equipped a Salvadorean air force fighter-bomber squadron, and about 60 equipped the 2nd Air Attack Squadron of Argentina's naval air arm.

Grumman F6F Hellcat

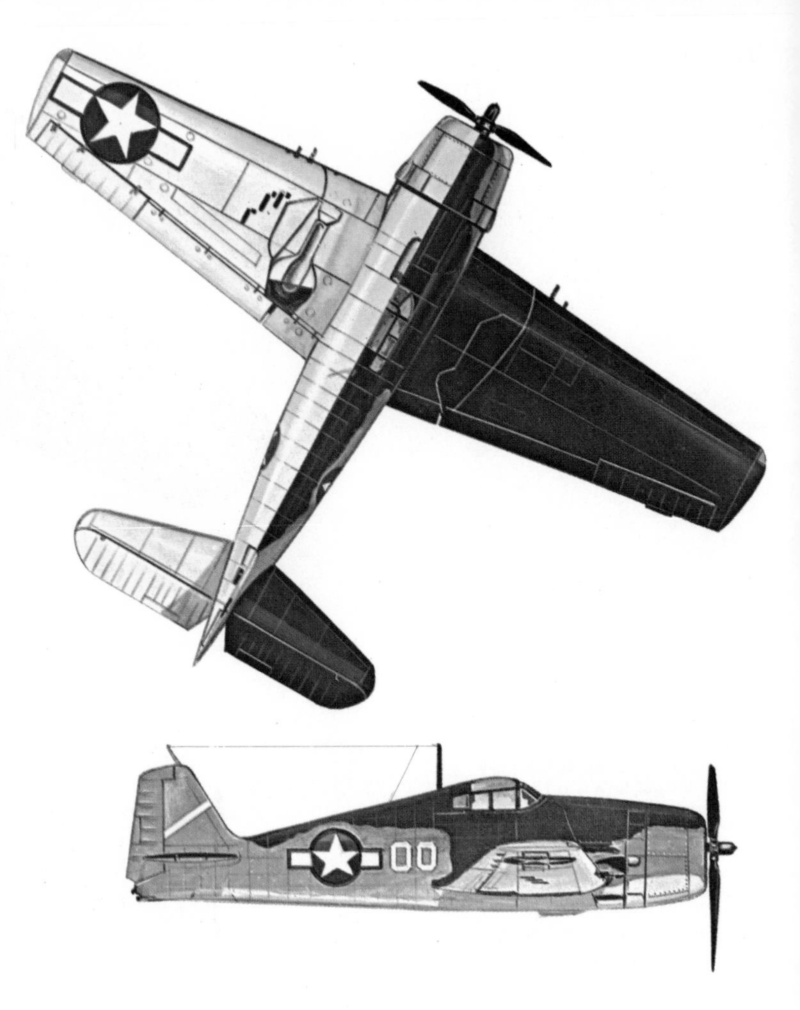

Grumman F6F-3 Hellcat of VF-9, US Navy, USS *Yorktown,* September 1943

Span: 42ft 10in (13.06m)
Length: 33ft 7in (10.24m)
Weight, normal take-off: 12,441lb (5,643kg)
Engine: 2,000hp Pratt & Whitney 12-2800-10 Double Wasp 18-cyl radial
Max speed: 375mph (604kmh) at 17,300ft (5,273m)
Operational ceiling: 37,300ft (11,370m)
Range: 1,090 miles (1,754km)
Armament: 6×0.50in Browning M-2 machine-guns

The Hellcat, essentially a larger and more powerful development of the F4F Wildcat, flew in its original XF6F-1 form on 26 June 1942, with a 1,700hp Wright R-2600-10 Cyclone engine. It was then re-engined with a 2,000hp Pratt & Whitney R-2800-10 Double Wasp to become the XF6F-3, flying in this form on 30 July 1942. Production F6F-3s were virtually unchanged from this aircraft. They began to appear early in October 1942, making their operational debuts with the FAA in July 1943, and with the USN in the attack on Marcus Island on 31 August by F6F-3s from USS *Essex, Yorktown* and *Independence.* The F6F rapidly replaced the Wildcat aboard USN attack carriers. FAA Hellcats participated in anti-shipping strikes off the Norwegian coast and in the attack on the *Tirpitz;* they were also operated extensively in the Far East. Production for the USN totalled 4,646 F6F-3s, including 18 F6F-3E and 205 F6F-3N night fighters; a further 252 were supplied to the FAA as the Hellcat Mk I, under Lend-Lease. Aerodynamic and control-surface improvements were introduced on the F6F-5, which entered production in 1944 and was able to operate in the fighter-bomber role with under wing weapons. The F6F-5 was powered by an R-2800-10W capable of 2,200hp using water-injection, and was both the principal and the last production Hellcat model. By November 1945, when production ended, 12,272 Hellcats had been manufactured. Of these, 6,436 were of the F6F-5 model, nearly a fifth of which were F6F-5N night fighters; and 930 others were essentially similar Hellcat Mk IIs for the FAA. Whereas its predecessor, the Wildcat, had been widely used in both the Atlantic and Pacific war areas, the Hellcat operated with the USN and the FAA predominantly in the Pacific, but was used aboard escort carriers in the Atlantic against U-boats. In service with land-based USMC units as well as carrier-based squadrons, it was officially credited with 4,947 victims, some 80 per cent of the 6,477 enemy aircraft destroyed in air-to-air combat by USN carrier pilots during the war, plus 209 by USMC pilots.

Yakovlev Yak-9

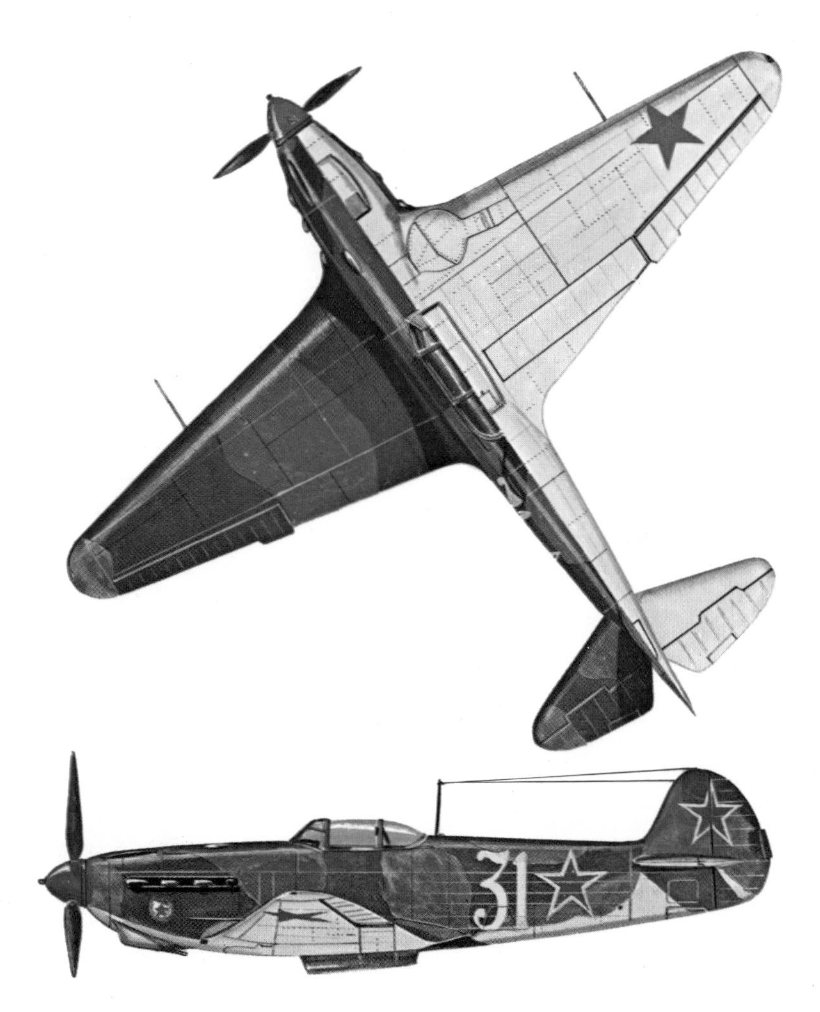

Yakovlev Yak-9D of a Soviet Air Force Guards Fighter Regiment, Crimea, spring 1944

Span: 32ft 9¾in (10.00m)
Length: 28ft 0⅝in (8.55m)
Weight, normal take-off: 6,867lb (3,115kg)
Engine: 1,210hp Klimov M-105 PF 12-cyl V-type
Max speed: 373mph (600kmh) at 11,485ft (3,500m)
Operational ceiling: 32,810ft (10,000m)
Range, normal: 808 miles (1,300km)
Armament: 1×20mm MPSh cannon; 1×12.7mm UBS machine-guns

The Yak-9, itself produced in a number of variants, represented the culmination of a highly successful line of single-engined fighters and trainers from the Yakovlev design bureau whose combined production total was some 30,000. It stemmed from the I-26 prototype of 1938, which became the Yak-1 in production in 1940, via the Yak-7, and the machines which acted as Yak-9 prototypes were originally designated Yak-7DI, signifying that they were designed as long-range fighters. They appeared in the first half of 1942, differing from the standard Yak-7B fighter chiefly in making greater use of light alloys. Production began in autumn 1942, and the Yak-9 was in operational service by the turn of the year in the Stalingrad fighting. In 1943 the Yak-9 began to be used as an anti-tank aircraft, being modified for this purpose as the Yak-9T to carry a 37mm cannon or a lighter weapon in the forward part of the fuselage. This was followed in 1944 by the Yak-9K, mounting a 45mm cannon that fired through the propeller shaft. The Yak-9B was a fighter-bomber version equipped to carry a 992lb (450kg) bomb internally, and in 1943-44 the Yak-9D and Yak-9DD emerged as variants with their range further increased to provide fighter cover for advancing troops and for bombing raids over enemy-held territory. One squadron of these, flying from southern Italy after the Italian armistice, provided support for the partisan forces in Yugoslavia, and other Yak-9 variants served with Polish and French units (including the celebrated Normandie-Niemen group) fighting in the USSR. The last major version to serve during the war was the all-metal Yak-9U, whose prototype flew in January 1944. This became operational during the second half of that year and was characterised chiefly by further aerodynamic refinements and the adoption of the new 1,600hp VK-107A engine which raised the top speed to 435mph (700kmh). The Yak-9U climbed from sea level to 16,400ft (5,000m) in nearly 30 seconds less than the Messerschmitt Bf 109G. The final Yak-9 variant (known briefly as the Yak-11) was the Yak-9P of 1945. This saw little service in World War 2, but was a standard postwar fighter and fighter-bomber with Soviet air forces, including the North Korean Air Force during 1950-53.

Boeing B-29 Superfortress

Boeing B-29 Superfortress of the 795th BS, 468th BG, US 20th Air Force, China-Burma-India theatre, early autumn 1944

Span: 141ft 3in (43.05m)
Length: 99ft 0in (30.18m)
Weight: 135,000lb (61,235kg)
Engines: 4× 2,200hp Wright R-3350-23 Cyclone 18-cyl radials
Max speed: 357mph (575kmh)
Operational ceiling: 33,600ft (10,240m)
Range with 10,000lb (4,536kg) bomb load: 3,250 miles (5,230km)
Armament: 1×20mm cannon; 10×0.50in machine-guns
Max bomb load: 20,000lb (9,072kg) internally

Design of the Superfortress began well before America's entry into World War 2, when the Boeing Model 345 was developed to a USAAC requirement of February 1940 for a 'hemisphere defense weapon'. In August 1940 two prototypes, designated XB-29, were ordered by the USAAF, and the first was flown on 21 September 1942. A much larger aeroplane than Boeing's earlier B-17, it had a circular-section, pressurised fuselage, remote controlled gun turrets and four 2,200hp Wright R-3350-13 Cyclone radial engines. By the time of the first flight, nearly 1,700 B-29s had been ordered. The first pre-production YB-29 Superfortress flew on 26 June 1943, and squadron deliveries began in the following month to the 58th BW. The first operational B-29 mission was carried out on 5 June 1944, and the first attack upon a target in Japan on 15 June 1944. It was during this month that the B-29s moved to the bases in the Marianas Islands, from whence they subsequently mounted a steadily increasing bombing campaign against Japan. Apart from the direct damage caused by this campaign, it was responsible for many Japanese aircraft from other Pacific battle fronts being withdrawn for home defence duties, although comparatively few types were capable of effective combat at the altitudes flown by the American bombers. B-29s also carried out extensive minelaying in Japanese waters; 118 others became F-13/F-13A photo-reconnaissance aircraft. Finally, two B-29s brought the war to its dramatic close with the dropping of atomic bombs on Hiroshima and Nagasaki on 6 and 9 August 1945 by *Enola Gay* and *Bockscar.* Shortly after VJ-day over 5,000 were cancelled, but when B-29 production ended early in 1946 the three Boeing factories had completed 2,756 B-29s and B-29As; in addition, 668 B-29s were manufactured by Bell, and 536 B-29s by Martin; 311 of the Bell machines were converted to B-29Bs with reduced armament. In 1945, an order was placed for 200 B-50s, improved B-29s with 3,500hp R-4360 Wasp Major engines, but cut to 60 after VJ-day. The principal external difference was the increased height of the fin; improvements included a lighter wing. In total 80 B-50As, 45 B-50Bs, 222 B-50Ds (with in-flight refuelling), and 27 TB-50H bomber trainers were built. Conversions included strategic reconnaissance aircraft for SAC, crew trainers, weather reconnaissance aircraft and tankers. The B-50 also serve with the RAF, as the Washington.

Hawker Tempest

Hawker Tempest Mk V Srs 1 (JN766) of No 486 Sqdn, RNZAF, UK, *circa* late spring 1944

Span: 41ft 0in (12.50m)
Length: 33ft 8in (10.26m)
Weight; normal take-off: 11,500lb (5,217kg)
Engine: 2,180hp Napier Sabre IIA, B or C 24 cyl H-type
Max speed: 436mph (701kmh) at 18,500ft (5,640m)
Operational ceiling: 36,500ft (11,125km)
Range, normal: 740 miles (1,191km)
Armament: 4×20mm Mk II cannon
Max bomb load: 2,000lb (907kg), or 8×60lb (27kg) RPs

The Tempest, originating as the Hawker P1012 in late 1941, was originally named Typhoon Mk II. Despite some similarity to the Typhoon Mk IB, the P1012 was a new design, developed specifically to overcome the Typhoon's performance limitations as an interceptor, mainly due to lack of structural integrity, its recalcitrant engine, poor performance above medium altitudes, and its wing's great thickness and relatively high aspect ratio, dictated by the requirement to accommodate 12 machine guns or four cannons. Camm designed a thin, elliptical wing, using more compact cannon, and rehoused the fuel tanks in the forward fuselage, thus lengthening it, but retained a tubular centre-section and monocoque rear fuselage. A dorsal fin fillet was later introduced. Two prototypes were ordered to Specification F10/41. The first flew on 2 September 1942 with a Napier Sabre IV, but the second had a Sabre II. Contracts for 400 Sabre IV-powered Tempest Mk Is were later amended to the 2,180hp Sabre II-engined version, which as the Mk V, was the first to enter production. (The Rolls-Royce Griffon-powered Mks III and IV were abandoned.) This was the only Tempest to see operational wartime service, designated Mk V Srs I with a Sabre IIA and Mk V Srs IIB with a Sabre IIB and fully buried guns; 805 Mk Vs were built. The first Tempests were delivered to No 3 Sqdn RAF and No 486 Sqdn RNZAF in April 1944, and flew many cross-Channel sorties before and after the invasion of Normandy. Soon after the invasion, they rapidly became one of the principal fighters employed to combat the V1 flying bombs over southern England, accounting for 638, more than one-third of those destroyed by the RAF. Eight RAF squadrons operated the Mk V during World War 2; it remained in RAF service until 1948. Pierre Clostermann called it 'a superb combat machine'. The Mk VI (142 built) was an improved model with Sabre V, but did not see service until after the war. The Mk II, chronologically the last serving version, powered by a 2,520hp Bristol Centaurus V or VI radial, had begun development in 1943 but its Centaurus met troubled development. It served postwar with three squadrons in India, one in the Far East, three in Germany, and two in the UK; 472 were built. The Mk II remained in RAF service until 1951 and with India and Pakistan until 1953. The Mks V and VI were relegated to target-towing duties in the early postwar period.

Messerschmitt Me 262 Schwalbe/Sturmvogel

Messerschmitt Me 262A-1a Schwalbe of 3/*JG* 7 *Nowotny, Luftwaffe,* Brandenburg, March 1945

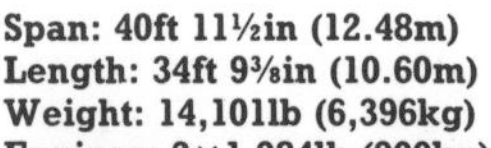

Span: 40ft 11½in (12.48m)
Length: 34ft 9⅜in (10.60m)
Weight: 14,101lb (6,396kg)
Engines: 2×1,984lb (900kg) st Junkers Jumo 109-004B-1 turbojets
Max speed: 541mph (870kmh) at 19,685ft (6,000m)
Operational ceiling: 37,565ft (11,450m)
Range: 652 miles (1,050km)
Armament: 4×30mm MK 108 cannon

Design of the Me 262 jet fighter, Messerschmitt project 1065, began in 1938, yet, due to delays in the development and delivery of satisfactory engines, the depredations of Allied air attacks, and Hitler's refusal to be advised regarding its most appropriate role, it was six years before it entered *Luftwaffe* service. Even then, only a fraction of those manufactured before VE-day became operational. A mock-up was completed during the latter half of 1939, and the RLM ordered three prototypes in spring 1940. These were completed long before their engines arrived and, to test the design, the Me 262V1 flew on 18 April 1941, with dummy jet-engine nacelles, and a 700hp Jumo 210 piston engine in the nose. On 25 March 1942 it attempted to fly with two underwing BMW 003 gas turbines, and the Jumo 210. The first all-jet flight was not made until 18 July 1942, by the third prototype powered by two 1,852lb (840kg) st Jumo 004 turbojets. Several prototypes were completed for trials with armament and equipment installations, and from the fifth onward tricycle gear replaced the tailwheel type. A pre-series batch of Me 262A-0s was completed in spring 1944, but air attacks upon Messerschmitt's Regensburg factory seriously affected plans for priority mass-production and delayed the planned introduction of the Me 262 into operational service until July 1941. The principal versions which became operational were the Me 262A-1A Schwalbe (Swallow) interceptor, built in a number of sub-types with alternative armament installations; and the Me 262A-2a Sturmvogel (Stormbird), produced as a result of Hitler's insistence upon developing the aircraft as a bomber, fitted with external bomb racks. Other variants built included the ground-attack Me 262A-3a, the PR A-5a, a tandem 2-seat trainer, designated Me 262B-1a, and one example was completed of the proposed B-2a 2-seat night fighter. The few Me 262C models completed were fitted with rocket motors in the fuselage to boost the climb to interception altitudes. Less than 600 Me 262s had been produced by the end of 1944, but by VE-day the total had risen to 1,433. Probably less than a quarter saw front-line combat service, and losses among these were quite heavy. In air-to-air combat, the Me 262 never engaged its British counterpart, the Meteor, but many were destroyed by Allied piston-engined Mustang, Thunderbolt, Spitfire and Tempest fighters.

Gloster Meteor

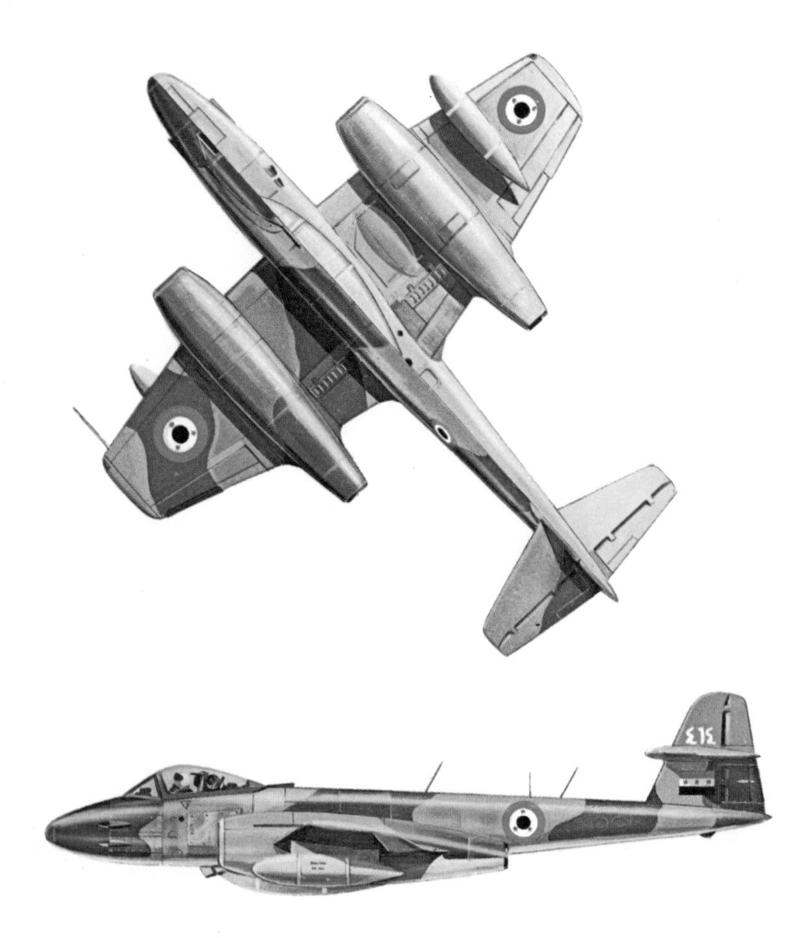

Gloster Meteor F Mk 8 of the Syrian Air Force, 1952-1956

Span: 37ft 2in (11.33m)
Length: 44ft 7in (13.59m)
Weight, normal take-off: 19,065lb (8,648kg)
Engines: 2×3,500lb (1,588kg) st Rolls-Royce Derwent 8 turbojets
Max speed: 592mph (953kmh) at sea level
Operational ceiling: 44,000ft (13,410m)
Range with auxiliary tanks: 1,090 miles (1,754km)
Armament: 4×20mm Hispano cannon; provision for underwing RPs

The Meteor was Britain's first production jet-propelled aircraft. Specification F9/40, issued in late 1940, was written around a Gloster proposal for a tricycle undercarriage, jet-fighter, with four cannons, and twin engines. Eight prototypes were built, the fifth being first to fly, on 5 March 1943, powered by de Havilland H.1 engines. Only 15 of the 20 Meteor F Mk Is built were delivered to the RAF, to No 616 Sqdn, making their operational debut against the V1s in July 1944. The F Mk III re-equipped No 616 Sqdn in December 1944, and in January 1945 began operating from the Continent. Used for ground attack, they were prohibited from over-flying enemy territory and aerial combat. Early F Mk IIIs had 2,000lb (910kg) st Wellands, later ones Rolls-Royce Derwent engines. The F Mk 4 had 3,000lb (1,360kg) st Derwent 5s, long-chord nacelles, and re-designed clipped wings: 535 were built, 168 being exported to Argentina, Holland, Egypt, Belgium and Denmark. A total of 712 T Mk 7 2-seat trainers were built, 72 being exported. The F Mk 8 had a longer fuselage, greater fuel load, ejector seat, square tail fin, and 3,500lb (1,600kg) st Derwent 8s. It was the main RAF day interceptor between 1950 and 1954; 1,183 were built, many for foreign air forces. From July 1951 an RAF Meteor squadron was operational in Korea. Unequal to the MiG-15, they operated in the ground attack role from January 1952. The fighter reconnaissance FR Mk 9 and unarmed PR Mk 10 with long-span wings, based on the F Mk 8, were operated by the RAF, Ecuador, Israel and Syria. The night-fighter NF Mk 11, first flown on 31 May 1950, was developed by Armstrong Whitworth to Specification F44/46, using the pressurised T Mk 7 2-seat cockpit, F Mk 8 tail, and PR Mk 10 wings; a longer nose housed the radar scanner. A total of 344 (including three prototypes) was built, 80 being sold to Belgium, Denmark and France. The NF Mk 12, the tropicalised NF Mk 13, and the NF Mk 14 were developed from the NF Mk 11 (total built 240), serving with the RAF, Egypt, France, Israel and Syria. The Meteor's standard armament was four 20mm cannon in the nose; for ground-attack it could carry two 1,000lb, or smaller bombs, or 16 95lb aerial rockets, underwing, under-fuselage fuel tank and two wing drop-tanks. A total of 3,875 Meteors, in 34 sub-types, was built, production ending in April 1954. The Meteor was retired from front-line RAF service in September 1961; a few remained for secondary duties. Ecuador was still operating F Mk 9s in front-line service until 1970.

Douglas A-26/B-26 Invader

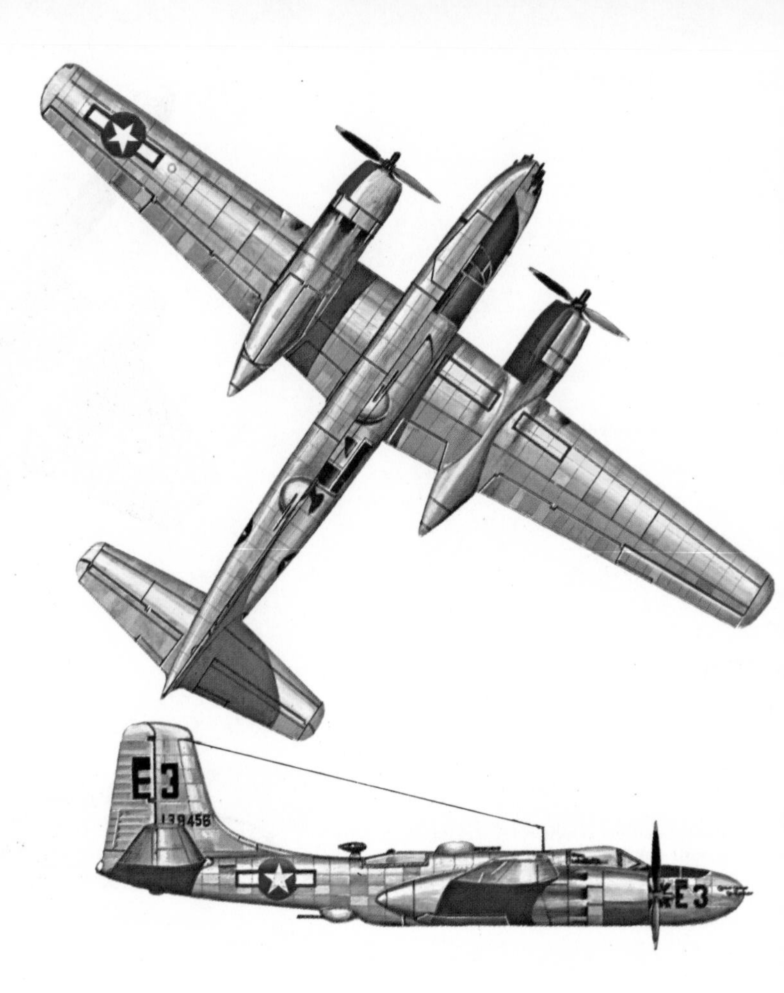

Douglas A-26B Invader of the US Ninth Air Force, Europe, 1944-45

Span: 70ft 0in (21.34m)
Length: 50ft 0in (15.24m)
Weight: 35,000lb (15,876kg)
Engines: 2× 2,000hp Pratt & Whitney R-2800-27 or -71 Double Wasp 18-cyl radials
Max speed: 355mph (571kmh) at 15,000ft (4,570m)
Operational ceiling: 22,100ft (6,735m)
Range: 1,400 miles (2,253km)
Armament: 10×0.50in machine-guns
Max bomb load: 4,000lb (1,814kg) internally

The Douglas Invader's design was begun in January 1941. It was initiated as a successor to the Douglas A-20 Havoc and designated in the 'A' for attack series. In June 1941 the USAAF ordered three prototypes each to a different configuration, the first flying on 10 January 1942. The XA-26 was an attack bomber, with a 3,000lb internal bomb load, twin guns, in a transparent nose and two each in dorsal and ventral turrets; the XA-26A night fighter had radar in a solid nose four cannon in a ventral tray and four 0.50in guns in a dorsal turret; the XA-26B had a short nose mounting a 75mm cannon. The initial production model, the A-26B, additionally had six 0.5in wing machine-guns, and the bomb load increased. Later batches introduced R-2800-79 engines with water injection, boosting power and performance at altitude, eight nose guns; and additional gun-packs, RPs or 2,000lb (907kg) of bombs could be carried beneath the wings. To concentrate fire-power, the dorsal guns could be locked forward and fired by the pilot. Douglas built 535 water-injection Bs, following the initial 825. The A-26 made its European operational debut in autumn 1944, and its first Pacific appearance early in 1945. The A-26C, appearing in 1945, saw limited war-service. Similar to the B, this had the twin-gun transparent 'bombardier' nose. After VJ-day, large numbers of orders were cancelled, but even so, 1,091 A-26Cs were completed. In Europe alone, Invaders flew over 11,000 sorties and dropped more than 18,000 tons of bombs for the loss of 67 aircraft in combat; curiously, they destroyed only seven enemy aircraft. Postwar, redesignated B-26 as a bomber after the Martin B-26 Marauder was withdrawn in 1958, they became a standard USAF type. In the Korean War the Invader proved ideal for close support and night intruder operations. Its usefulness in limited war engagements was emphasised in Vietnam, and On Mark Engineering brought 40 B-26Bs up to B-26K standard for the COIN role. Modifications included installing higher-powered engines and airframe strengthening to permit carriage of a 6,000lb ordnance load on eight underwing points, in addition to the internal load. Until the late 1960s Invaders, either Bs or Cs, though not first-line equipment, served with the French, Indonesian and Turkish air forces, but were operated in their original roles by Brazil, Chile, the Dominican Republic, Guatemala, Peru, Saudi Arabia and South Vietnam.

Douglas AD/A-1 Skyraider

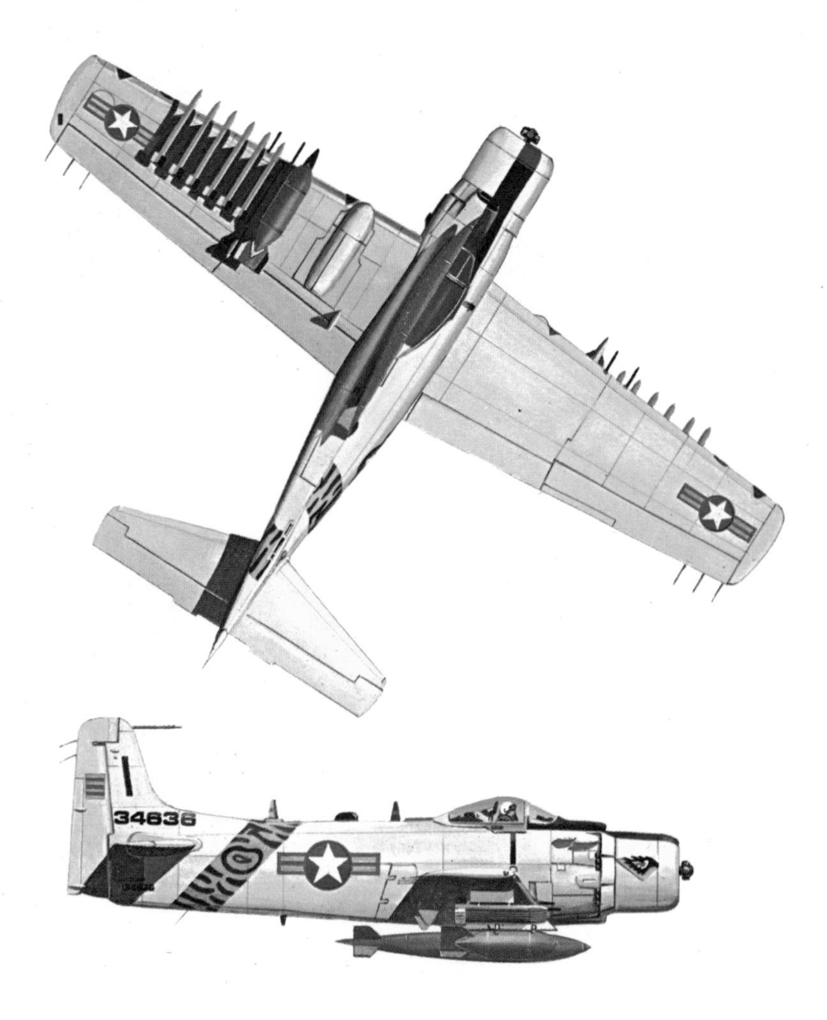

Douglas A-1H Skyraider of the Vietnam Air Force, 1965

Span: 50ft 0in (15.24m)
Length: 39ft 2½in (11.95m)
Weight: 25,000lb (11,340kg)
Engine: 2,700hp Wright R-3350-26WA 18-cyl radial
Max speed: 285mph (459kmh) at 18,500ft (5,640m)
Operational ceiling: 28,500ft (8,685m)
Radius on internal fuel: 1,145 miles (1,840km)
Armament: 4×20mm cannon
Max bomb load: 8,000lb (3,630kg) externally

A substantial and simplified redesign of Ed Heinemann's XSB2D-1 Destroyer of 1941, of which 28 were built as the BTD-1, the AD-1 Skyraider was first flown as the XBT2D-1 Destroyer II on 18 March 1945. The Skyraider had greatness thrust upon it. Production orders were not cancelled, unlike orders for other aircraft, but merely cut after VJ-day, and moreover it was due for retirement after five years service with the USN when the Korean War broke out in 1950. Its record during the three year conflict, and its amazing versatility, kept it in continued production until 1957 when, after a 12-year run, the Douglas factories had turned out 3,180 Skyraiders. Rather more than half of these were accounted for by the first five models (including one 1,051 A-1Ds), which had passed out of front-line service by the mid-1960s; the remainder comprised 670 A-1Es, 713 A-1Hs, and 72 A-1Js. The A-1E was primarily a 2-seat attack bomber, but capable of being converted by a ready-made kit system to night attack, early warning, ECM (EA-1E), 12-seat transport, 4-litter ambulance, 2,000lb (907kg) freighter, photographic reconnaissance or target towing duties. The A-1H was a single-seat low-altitude attack bomber, the A-1J being similar except for a higher-powered engine and an airframe strengthened to carry a tremendous variety of loads. The Skyraider's load-carrying capacity was one of its most remarkable features, and played a major part in the success of this aircraft under operational conditions. The specification to which it was designed called for a load of 1,000lb (454kg), but Skyraiders regularly flew with more than 8,000lb (3,630kg) of external ordnance, and the type has been flown with more than 14,000lb (6,350kg) of under wing stores – greater than the aircraft's basic empty weight. The Skyraider's variety of load was equally impressive, and was augmented by four 20mm wing cannon. Skyraiders had been withdrawn from USN service by the end of 1967, but in the early 1970s the USAF operated several squadrons in Vietnam, and the South Vietnamese Air Force received over 100 A-1Es and A-1Hs. Other Skyraiders in the Far East at that time included about a dozen A-1Ds, resold by France in 1965 to the air force of Cambodia. Ex-French A-1Ds were also supplied to the Republic of Chad (five) and the Central African Republic (ten).

Lockheed P2V/P-2 Neptune

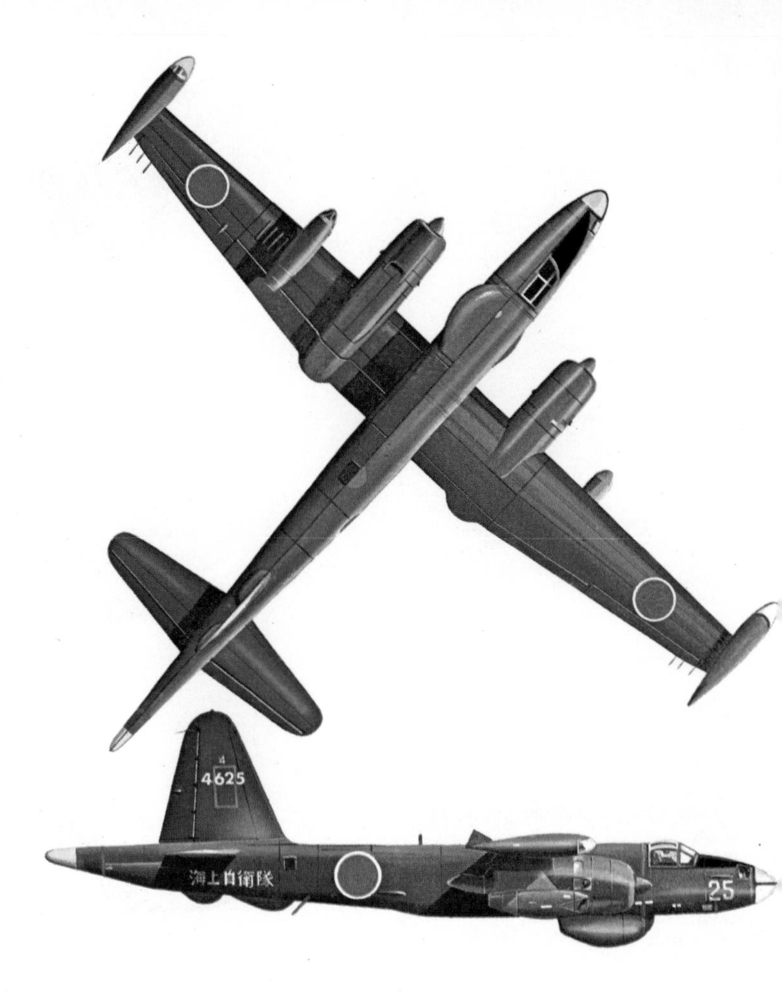

Kawasaki-built Lockheed P-2H (P2V-7) of the JMSDF, *circa* 1962

Span (over tip-tanks): 103ft 10in (31.65m)
Length: 91ft 8in (27.94m)
Weight: 79,895lb (36,240kg)
Engines: 2×3,500hp R-3350-32W Turbo-Compounds, and 2×3,400lb (1,542kg) st Westinghouse J34-WE-36 auxiliary turbojets
Max speed: 356mph (573kmh) at 10,000ft (3,050m)
Operational ceiling: 22,000ft (6,700m)
Range: 2,200 miles (3,540km)
Armament: 2×0.5in machine-guns
Max bomb load: None

A continuous/service record of nearly 30 years is a measure of the Neptune's place in the field of patrol aircraft. The first steps towards the Neptune's design were taken before Pearl Harbor, although the prototype XP2V-1 did not fly until 17 May 1945. Intended basically for anti-shipping and submarine patrol, it has fulfilled secondary torpedo attack, minelaying and reconnaissance roles. Entering USN service in March 1947, the early P-2A to D (originally P2V-1 to -4) accounted for 233 aircraft including two prototypes; they were replaced by the P-2E, F and H. The P-2E (P2V-5, first flight 29 December 1950) gradually introduced such features as a redesigned glazed nose, MAD in the tail, and more modern anti-submarine systems. The P-2F, first flown on 16 October 1952, introduced a revised nose, tip-tanks, ventral radome and wider load-carrying capabilities. The P-2H (P2V-7, first flight 26 April 1954) had further changes and additional power from two pod-mounted jet engines beneath the wings, the latter retrospectively applied to all P-2Es in service, and to some P-2Fs which were then redesignated P-2G. The largest Neptune operator outside the USA was France's *Aéronavale* with six squadrons of P-2E/F/H, reduced to two by 1980. Neptunes served with the RCAF (squadron of 25 P-2H); the *Koninklijke Marine* (four squadrons of P-2E/H); the *Fôrça Aérea Portuguesa* (squadron of 12 P-2E); the RAAF (squadron of 12 P-2H); the JMSDF (five squadrons of P-2H – 76 aircraft, of which 48 were licence-built by Kawasaki); Argentina's *Aviación Naval* (squadron of six P-2E); and the *Fôrça Aérea Brasileira* (squadron of 14 P-2E). Two missile-carrying variants were the USN's MP-2F, with the anti-submarine Petrel, and the Dutch P-2H with the Nord AS12. US production amounted to 1,099 aircraft, including 424 P-2Es, 83 P-2Fs and 359 P-2Hs.

To replace the JMSDF's P-2Hs Kawasaki developed the improved P-2J from the P-2H, with a 4ft 2in (1.27m) longer fuselage, redesigned main landing gear and larger rudder. Licence-built General Electric turboprops replaced the piston engines, and the auxiliary turbojets were of Japanese design. Operational equipment is comparable to the Lockheed P-3 Orion's. A prototype flew on 21 July 1966, and a production example on 8 August 1969. It entered service in February 1971; 83 were built.

Republic F-84 Thunderjet

Republic F-84G Thunderjet of the Royal Thai Air Force, *circa* 1959

Span: 37ft 5in (11.40m)
Length: 38ft 5in (11.71m)
Weight: 23,525lb (10,671kg)
Engine: 5,600lb (2,540kg) st Allison J35-A-29 turbojet
Max speed: 622mph (1,001kmh) at sea level
Operational ceiling: 40,500ft (12,345m)
Range: 1,470 miles (2,366km)
Armament: 6×0.50in machine-guns
Max bomb load: 2×1,000lb (454kg) or 32×5in RPs externally

Design of the F-84, beginning in late 1944, was completed in February 1945. The XP-84 first prototype flew on 28 February 1946, powered by a 3,750lb (1,700kg) at YJ35 turbojet; an improved 4,000lb (1,815kg) st J35 powered the second. Republic built 15 differing pre-production YP-84s in 1946-47, powered by J35-15s. They were used as development aircraft, proving the six 0.50in M-2 machine-gun armament, underwing points for bombs or tanks, tip-tanks, ejection seat, and the pressurised, air-conditioned cockpit which were incorporated in the initial production series, the F-84B (the designation became F-84 in June 1948). Powered by a 4,000lb (1,815kg) st J35-A-15, 226 were built; from the 86th eight underwing zero-length retractable launchers for 32 5in RPs were introduced. The F-84C (191 built) had a higher loaded weight, sequencer for multiple bomb or rocket loads, and other modifications and the J35-A-13. The greatly improved F-84D (153 built) had the 5,000lb (2,270kg) st J35-A-17D and a restressed, reinforced wing, and other improvements. The F-84E (843 built) had a lengthened fuselage and was restressed throughout. The final version, the F-84G, was developed from the F-84E, with considerable structural strengthening, and was powered by the 5,600lb (2,540kg) st J35-A-29. A total of 3,025 was built, production ending in July 1953. The F-84G had no in-built extra fuel capacity but had Boeing 'flying-boom' in-flight refuelling and an auto-pilot to allow rapid global mobility and theatre reinforcement. It was operated by the USAF's TAC and ADC. Under MAP, 1,936 F-84Gs were assigned to Belgium, Denmark, France, Greece, Italy, the Netherlands, Norway, Portugal and Turkey; others were supplied to Iran, Yugoslavia, the Chinese People's Republic, and Thailand. The F-84G was flown by the USAF's Thunderbirds aerobatic team. It had 32 5in RPs in underwing launchers plus the standard six machine-guns. F-84s began flying combat missions in Korea in autumn 1950. Never a successful air superiority fighter, it had a low ceiling and low rate of climb and was inferior to the MiG-15. Moreover, its poor short-field performance was a hindrance in Korea, Europe and hot countries; the F-84E, however, introduced jet-assisted take-off (JATO). However, it had long range and high load carrying ability. The J35 powered all 4,457 F-84 series aircraft built; its poor thrust/weight ratio resulted in the F-84 becoming less competitive. They were, however, tough and were progressively developed, and quickly produced, serving NATO through the Cold War period. The F-84 served with Yugoslavia until 1974 and Portugal in Angola until 1975.

North American F-86 Sabre

North American F-86K-10-NA Sabre of No 337 Sqdn, Royal Norwegian Air Force, *circa* 1960

Span: 37ft 1½in (11.32m)
Length: 40ft 11in (12.47m)
Weight: 19,952lb (9,050kg)
Engines: 5,425/7,500lb (2,460/3,402kg) st General Electric J47-GE-17B afterburning turbojet
Max speed: 612mph (985kmh) above 36,000ft (11,000m)
Operational ceiling: 49,600ft (15,120m)
Range with external tanks: 745 miles (1,200km)
Armament: 4×20mm cannon

The XP-86 prototype flew on 1 October 1947, powered by a 4,000lb (1,814kg) st General Electric J47. The first production F-86A flew in May 1948, powered by a 5,200lb (2,359kg) J47; 554 F-86As were built, entering service in early 1949. It was followed in late 1950 by the F-86E, with an 'all-flying' tail. The F-86B and C were experimental. The F-86F, the last day fighter variant, was developed from the E, having a more powerful J47 and structural redesign. Originally a day fighter, with six 0.50in machine-guns, the F-86 proved itself in Korea, where it had an 11:1 'kill' ratio over the MiG-15, but was also extensively employed as a fighter-bomber. The F-86, principally the F, served over 20 NATO and friendly air forces. The 'open nose' early variants were licence-built by Canadair (Orenda engines), by CAC in Australia (Rolls-Royce Avons), and by Mitsubishi in Japan. All-weather interceptor version. Development of the F-86D, began early in 1949, and a YF-86D prototype (one of two built) flew on 22 December 1949. The F-86D was unusual in being single-seat, when contemporary practice dictated a second crew member for radar and electronics. Problems with this equipment kept the F-86D from full effectiveness until mid-1953, although it had entered USAF service in 1951. The F-86D had virtually a new fuselage, embodying a 'chin' intake under a nose radome, and a retractable ventral pack of 24 2.75in rockets replacing a gun armament. From 1958 the US dispersed many of its F-86Ds to other countries, recipients including Nationalist China, Denmark, Greece, Japan, South Korea, the Philippines, Turkey, and Yugoslavia. A total of 2,504 F-86Ds were built – the largest production of any Sabre; 981 were converted to extended-span F-86Ls, with more advanced avionics, to serve as a part of America's SAGE (Semi-Automatic Ground Environment) defence system. Some F-86Ls were delivered to Thailand. An export version of the D, evolved for NATO, was the F-86K, which had four 20mm M24A-1 cannon and two Sidewinders to replace the rockets. F-86K production totalled 341, between 1954 and 1958. North American built 120, the remainder being assembled by Fiat in Italy, the final 45 being completed with the extended-span wings, with modified leading-edges, of the L. American-built F-86Ks were supplied to Norway and Denmark. Italian-built examples were delivered to the Italian Air Force, France, Germany, Netherlands, and Norway. The 74 surviving Luftwaffe F-86Ks were sold to Venezuela in 1966; the Dutch aircraft were transferred to Turkey.

MiG-15 (Fagot/Midget)

SB LIM-1 (Polish licence-built Mikoyan/Gurevich MiG-15UTI) of the Polish Air Force, *circa* 1958

Span: 33ft 2¼in (10.11m)
Length: 36ft 3½in (11.06m)
Weight, normal take-off: 10,891lb (4,940kg)
Engine: 5,952lb (2,700kg) st Klimov RD-45 turbojets
Max speed: 628mph (1,010kmh) at sea level
Operational ceiling: 50,855ft (15,500m)
Range, with drop-tanks: 628 miles (1,010km)
Armament: 1×23mm NS-23 cannon or 1×12.7mm BS machine-gun
Max bomb load: None

Winner of three competing designs to a 1946 fighter specification, the I-310 prototype of the MiG-15 powered by a Rolls-Royce Nene engine, first flew on 2 July 1947. It was designed by Artem Mikoyan and Mikhail Gurevich, who since the outbreak of World War 2 had led the MiG design bureau. Its wings, with anhedral and 35 degrees of sweep, combined German and Russian swept-wing research. Initial production aircraft, which entered service in late 1948, were powered by a Klimov RD-45 or VK-1 turbojet, developed in Russia from examples of the Rolls-Royce Nene which, innocently, Britain had sold to the USSR in 1946. Initial armament was two 23mm nose cannon; a 37mm cannon was added later. The MiG-15, whatever its design shortcomings, deserves its place in aviation history as the first jet fighter to enter wide-scale service in the USSR and its satellites; and for the sheer numbers in which it was built, which some authorities put as high as 18,000 – a record for jet aircraft production. It was designed for easy mass production and maintenance, ideal for export to the USSR's less-developed satellites. It was licence-built in Czechoslovakia as the S-102, in Poland as the LiM-1, and in China. MiG-15 losses in the Korean War were 11 times those of its Allied counterpart, the North American F-86 Sabre, due as much to the inexperience of their Chinese and North Korean pilots as to technical weaknesses (those flown by Russian 'technical advisers' were noticeably better opponents). It introduced many air forces to the age of jet flying. The major variants were the MiG-15S and SD – the major production fighters; the MiG-15T and *bis*I target tugs; the MiG-15R and *bis*R reconnaissance fighters; the MiG-15P all-weather fighter; the MiG-15SB fighter-bomber; and the MiG-15UTI 2-seat trainer. Although the MiG-15 (NATO Fagot) has long been superseded in its front-line capacity, many countries – including China, Cuba, Indonesia, Morocco, Syria and the United Arab Republic – used the MiG-15UTI (Nato Midget) into the late 1960s, and some are still serving in the early 1980s. In the UTI version the basic MiG-15 fuselage is adapted to accommodate a second crew position in tandem beneath a elongated canopy, the fuel capacity thus displaced being made up by slipper tanks on the outer wing panels. The first MiG-15UTI was flown in 1949, entering Soviet Air Force service three years later as a replacement for the Yak-17UTI. It retained the two 23mm Nudlelmann cannon of the early fighters, and was used for operational conversion, but latterly for advanced flying and armament training.

Ilyushin Il-28 (Beagle)

Ilyushin Il-28 (Beagle) of the German Democratic Republic *Luftstreitkräfte, circa* 1962

Span (excluding tip-tanks): 70ft 4½in (21.45m)
Length: 57ft 11in (17.65m)
Weight, normal take-off: 40,565lb (18,400kg)
Engines: 2×5,952lb (2,700kg) st Klimov VK-1A turbojets
Max speed: 560mph (902kmh) at 14,760ft (4,500m)
Operational ceiling: 40,350ft (12,300m)
Range with max fuel: 1,490 miles (2,400km)
Armament: 4×23mm NR-23 cannon
Max bomb load: 6,614lb (3,000kg); normal: 2,205lb (1,000kg)

The Il-28 three-seat tactical light bomber is generally looked upon as the 'Russian Canberra', chiefly because the two were roughly contemporary designs intended to fulfil broadly similar functions. There is, however, no physical resemblance, and the Il-28 did not, like the British machine, rely on altitude and speed alone to take it to and from its target unmolested, carrying two 23mm cannon in the nose and two more in an Il-K-6 tail turret for defensive purposes. The Il-28 is of all-metal construction with a semi-monocoque fuselage but is much slimmer than the Canberra, and can carry a greater bomb load up to 6,600lb (3,000kg) within its short, deep bomb bay. It has, like the Canberra, a tricycle undercarriage. First flown on 8 August 1948, it was publicly demonstrated on May Day 1950. It first entered service with the V-VSs *Frontovaya Aviatsiya* in 1949, and, although obsolete, served into the late 1960s with this force and the Soviet Navy's A-VMF, as the Il-28T torpedo-bomber, despite the arrival of the Yak-28 Brewer. It was the first Soviet jet bomber to enter service and was the standard equipment of the Soviet light bomber units throughout the 1950s. Soviet production, ending in spring 1957, exceeded 5,000, and it was one of the basic types used to re-equip many pro-Communist air forces in the 1950s and early 1960s. The Il-28R was a tactical reconnaissance variant. The Il-28 equipped the air forces of Afghanistan, the Chinese People's Republic (which had 300-400 in service in 1980 and still builds the aircraft under the licence as the Harbin O-5), Czechoslovakia, the German Democratic Republic, Hungary, Indonesia (which had 25), the North Korea, Poland (which had one regiment), Romania (one regiment) and the United Arab Republic (four squadrons). A number of ex-UAR Il-28s were handed on to the Algerian Air Force in 1964. The total number in service outside Russia was estimated in the early 1970s at about 750. A conversion trainer variant, the Il-28U (Mascot) has a much revised solid, stepped nose with a second cockpit in tandem, and equipped most Il-28 units in small numbers. Other examples with detail variations have been employed by some air forces for radar training or meteorological research. *Aeroflot* briefly operated a civil version for mail-flights, designated Il-20, and others were converted for meteorological duties.

BAC (English Electric) Canberra

BAC (English Electric) Canberra B(I) Mk 8 of No 16 Sqdn, 2nd TAF, RAF, *circa* 1964-65

Span: 63ft 11½in (19.49m)
Length: 65ft 6in (19.96m)
Weight, normal take-off: 50,992lb (23,130kg)
Engines: 2×7,500lb (3,400kg) st Rolls-Royce Avon Mk 109 turbojets
Max speed: 580mph (933kmh) at 30,000ft (9,145m)
Operational ceiling: 48,000ft (14,630m)
Range on internal fuel: 800 miles (1,287km)
Armament: 4×20mm Hispano cannon
Max bomb load: 3,000lb (1,360kg) internally; 2,000lb (907kg) or 2× ASMs externally

The Canberra, indisputably among the greatest postwar British military aircraft successes, served the RAF well for over two decades and equipped the air forces of eight other nations. Designed by William Petter to Specification B3/45, as a high-altitude bomber, the first prototype aircraft flew on 13 May 1949, with two Rolls-Royce Avons; the second prototype, as a precaution, had Nenes. The B Mk 1's radar-bombing system did not materialise, and the visual-bombing B Mk 2 (Avons) was the first production model, entering Bomber Command service in May 1951, soon followed by its reconnaissance counterpart, the PR Mk 3. These were supplanted respectively by the B Mk 6 and PR Mk 7, which had uprated Avon engines and integral wing tanks. The Canberra's normal internal bomb load is 6,000lb, but in the B(I) Mk 8, an intruder version for the 2nd TAF, this was halved to allow a ventral pack of four 20mm cannon to be installed beneath the bomb bay, external loads of rocket pods or two 1,000lb bombs being added underwing. The B(I) Mk 8 intruder has a two-man crew and a blister cockpit hood offset to port; the B(I) Mk 12 is similar. The PR Mk 9, has a B(I) Mk 8-style canopy and extended span for higher altitude operation. The RAF operated Canberras in nuclear strike and reconnaissance roles, at home, in Europe, and in the Near, Middle and Far East. Most B Mk 6s were converted to B Mks 15 or 16, unarmed bombers with full internal load plus the B(I) Mk 6's external capacity; these could carry Nord AS30 missiles externally. Bomber variants still serve with Argentina (Mk 62), Ecuador (Mk 6), France (Mk 6), India (B(I) Mk 58, B Mk 66), Peru (B Mks 2/78), South Africa (B(I) Mk 12), Venezuela (B Mk 2, B(I) Mks 8/82/88), and Zimbabwe (B Mk 2); India also operates the PR Mk 57 and Venezuela the PR Mks 3/83. A total of 1,376 was built in Britain in 22 variants, and under licence abroad. Martin acquired a US production licence in 1951, but undertook extensive redesigns. The main production model was the B-57B night intruder (202 built), with tandem seats underwing pylons, wing guns and a rotating bomb-bay, followed by 38 similar dual-control TB-57C and 68 similar target-towing B-57E. The 20 photographic and electronic reconnaissance RB-57Ds had a 106ft span; the General Dynamics strategic and weather reconnaissance RB-57F (21 converted) has a 122ft 6in span, revised tail surfaces, turbofans, and two auxiliary underwing turbojets; the B-57G (16 converted) was a night sensor version used in Vietnam.

De Havilland Venom

De Havilland Venom FB Mk 50 of the *Schweizerischen Flieger-Regimenter,* Swiss Air Force, *circa* 1964

Span: 41ft 8in (12.70m)
Length: 32ft 1⅞in (9.8m)
Weight, normal take-off: 13,650lb (6,190kg)
Engine: 4,850lb (2,200kg) st de Havilland Ghost Mk 103
Max speed: 584mph (940kmh) at sea level
Operational ceiling: 49,200ft (15,000m)
Range with drop-tanks: 1,045 miles (1,680km)
Armament: 4×20mm cannon; provision for 8×60lb RPs underwing
Max bomb load: 2,000lb (907kg) externally

The Venom was a development of the DH100 Vampire. The Vampire F Mk 1 entered RAF service in April 1945, followed by the F Mk 3 with greater fuel capacity and revised tail. The FB Mk 5, the first ground attack variant, had restressed, clipped wings with bomb pylons, entering service in 1948; the export FB Mk 6 had an uprated engine. The FB Mk 9 had a tropicalised cockpit, entering RAF service in 1952. The first 2-seat version, the RAF's NF Mk 10, served between 1950-52. The DH115 T Mk 11 2-seat advanced trainer served the RAF between 1952-1962; the Sea Vampire T Mk 22 was the FAA counterpart, and the T Mk 55 the export version. Large numbers served some dozen nations, and were licence-built in Switzerland (FB Mk 6), India and Australia. Standard powerplant was the de Havilland Goblin. To exploit their new Ghost engine, de Havilland developed a thinner wing with 17 degrees leading edge sweep. It was fitted to an FB Mk 5, which flew on 2 September 1950. Designated DH112 Venom FB Mk 1, de Havilland built 60 for the RAF. The NF Mk 2 was basically a 2-seat FB Mk 1. The NF Mk 3 (129 built) had improved handling, AI radar, tail surfaces, and clear-view canopy, and introduced ejector seats. NF Mk 2s brought up to NF Mk 3 standard were re-designated NF Mk 2A. The FB Mk 4, an FB Mk 1 development first flown on 29 December 1953, had powered ailerons, revised wings and NF Mk 3 tail; the 4,850lb (2,200kg) st Ghost 103 powered initial examples, superseded by the 5,300lb (2,400kg) st Ghost 105. Export versions included the FB Mk 50, basically an FB Mk 1, flown on 2 September 1949, for Switzerland, Italy and Iraq, and the NF Mk 51 for Sweden. Switzerland licence-built 150 FB Mk 50s, many still in service. Venezuela bought 22 FB Mk 4s. The RAF employed FB Mks 1 and 4 single-seat fighter-bombers and NF Mks 2 and 3 2-seat night-fighters in Germany, and the Middle and Far East until 1962. A score of nations used the Venom, some serving until the 1970s. The Sea Venom 2-seat, all-weather carrier-borne fighter met RN Specification N107. The FAA operated them until the late 1960s. The FAW Mk 20 had the 4,850lb st Ghost 103R, the FAW Mk 21, the 4,850lb st Ghost 104; and the FAW Mk 22, the Ghost 105. Australia bought 39 Sea Venom FAW Mk 53s, an export FAW Mk 21, serving on HMAS *Melbourne* between 1956-67. Sud-Aviation built 94 Sea Venoms as the Sud-Est Aquilon, with Fiat-built Ghost 48s, equipping two all-weather, carrier-borne *Aéronavale* squadrons until the late 1960s. The 202, unlike the original Aquilon 20, had an ejector seat; the 203 and 204 were single-seat and training versions respectively.

MiG-17 (Fresco)

Mikoyan-Gurevich MiG-17PF (Fresco-D) of the Republic of Indonesia Air Force, *circa* 1960

Span: 31ft 6¾in (9.62m)
Length: 37ft 8¾in (11.50m)
Weight, normal take-off: 12,390lb (5,620kg)
Engine: 5,732/7,452lb (2,600/3,380kg) st Klimov VK-1F afterburning turbojet
Max speed: 693mph (1,115kmh) at 9,845ft (3,000m)
Operational ceiling: 54,460ft (16,600m)
Range with underwing tanks: approx 1,285 miles (2,070km)
Armament: 3×23mm NR-23 cannon
Max bomb load: 500kg (1,102lb) externally (bombs, rockets or ASMs)

In January 1950 Artem Mikoyan recorded the first flight of the MiG-17, derived from his famous MiG-15 fighter, but it did not enter Soviet Air Force service until 1953, the latter half of 1952. Subsequently, it was built in Russia, China, Czechoslovakia (as the C-105) and Poland (LiM-5) on a scale which may have exceeded that of the MiG-15, and served with more than a score of countries for a remarkably long period; it was particularly active at the time of the Suez crisis in 1956. Most of these air arms re-equipped with supersonic MiG-21s, but MiG-17s could still be found in the early 1970s in the inventories of many air forces, particularly in Asia, the Middle East and Africa. Fresco-A was the initial production MiG-17, a day fighter with a 5,952lb (2,700kg) st non-afterburning VK-1A and armament of two 23mm and one 37mm cannon; few remained in service by the mid-1960s. The MiG-17P Fresco-B was its limited all-weather interceptor counterpart, distinguished by its 'lipped' air intake and small centrebody radar; small numbers were built. Most Fresco operators received either the MiG-17F (Fresco-C) day fighter or MiG-17PF (Fresco-D) limited all-weather and night fighter. The MiG-17F was the major production version, also built under licence in Poland as the LiM-5P, Czechoslovakia as the S-104 and China as the Shenyang F-5. A multi-purpose day fighter, it had a fixed armament of three 23mm NR-23 cannon with provision for bombs, unguided rockets or drop-tanks underwing. It was powered by the VK-1F afterburning version of the Klimov engine, and had larger airbrakes. The MiG-17PF also appeared as the MiG-17PFU (Fresco-E), with its cannon deleted in favour of four Alkali AAMs (a feature found on some -Ds). Polish designers evolved a 'battlefield' version of the MiG-17F, designated LiM-5M, with deeper, longer-chord inboard wing panels housing extra fuel and twin-wheel main landing gear units with low-pressure tyres. A limited number, produced by converting LiM-5Ps, served with the Polish Air Force in the late 1960s, but preference was given to a less complicated conversion, the LiM-6, to which standard many earlier Polish MiG-17s were modified. A photographic version was designated LiM-6R.

Republic F-84F Thunderstreak

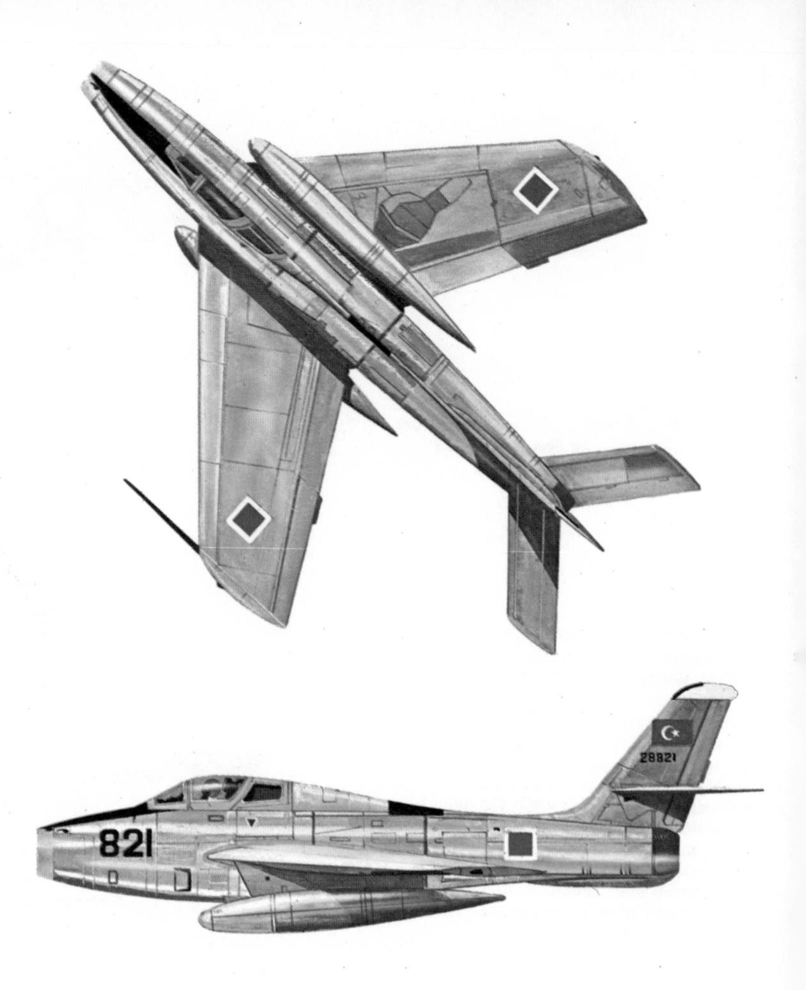

Republic F-84F Thunderstreak of the Turkish Air Force, *circa* 1964

Span: 33ft 7¼in (10.24m)
Length: 43ft 4¾in (13.23m)
Weight, normal take-off: 19,340lb (8,772kg)
Engine: 7,220lb (3,275kg) st Wright J65-W-3 turbojet
Max speed: 658mph (1,059kmh) at 20,000ft (6,100m)
Operational ceiling: 46,000ft (14,020m)
Radius with drop-tanks: 810 miles (1,304km)
Armament: 6×0.50in machine-guns
Max bomb load: 6,000lb (2,722kg) externally

Originally designated YF-96A and first flying on 3 June 1950, the Thunderstreak was a company-sponsored swept-wing development of the F-84G Thunderjet. Official interest, lukewarm at first, hardened soon after the outbreak of the Korean War and the type was ordered into production with a designation in the F-84 series and various design improvements, among them the substitution of the Wright J65 (Sapphire) in place of the original underpowered Allison. By August 1957, Republic and General Motors had built 1,410 F-84Fs for TAC and SAC (USAF) and 1,301 for various NATO air forces. The US Thunderstreaks were relegated to units of the ANG in the early 1960s and the *Armée de l'Air* and the German *Luftwaffe* phased out theirs in favour of the Mirage III or F-104G Starfighter in the mid-1960s. They remained in service with the Belgian, Greek, Italian, Dutch and Turkish air forces until about 1970. The Thunderstreak was most often to be seen with two underwing drop-tanks, which nearly doubled its fuel load, but its four wing strong-points were capable of lifting over 20 different combinations of offensive stores to a total weight of 6,000lb. It carried four 0.50in guns in the nose, and one in each wing root. The Thunderstreak's basic roles were interception or ground attack, but its excellent range also made it a useful escort aircraft. The RF-84F Thunderflash was a variant of the F-84F produced for photographic reconnaissance. Since the nose is the natural site for the camera bay in a single-engined aircraft, it was necessary to devise twin intake ducts located in the wing roots standing proud of the leading edge. Otherwise it had similar construction to its stablemate, and had only marginally less performance. The nose bay accommodated six cameras, and with underwing magnesium flare cartridges it could undertake night as well as day photographic missions. Like the Thunderstreak, its range was considerably extended by twin drop-tanks, and the Thunderflash had four 0.50in guns. Of 715 RF-84Fs built, 386 were supplied to Belgium, Nationalist China, Denmark, France, West Germany, Greece, Italy, the Netherlands, Norway and Turkey. The Dutch aircraft were replaced by the F-104 Starfighter. ANG units in the USA also employed the RF-84F.

Hawker Hunter

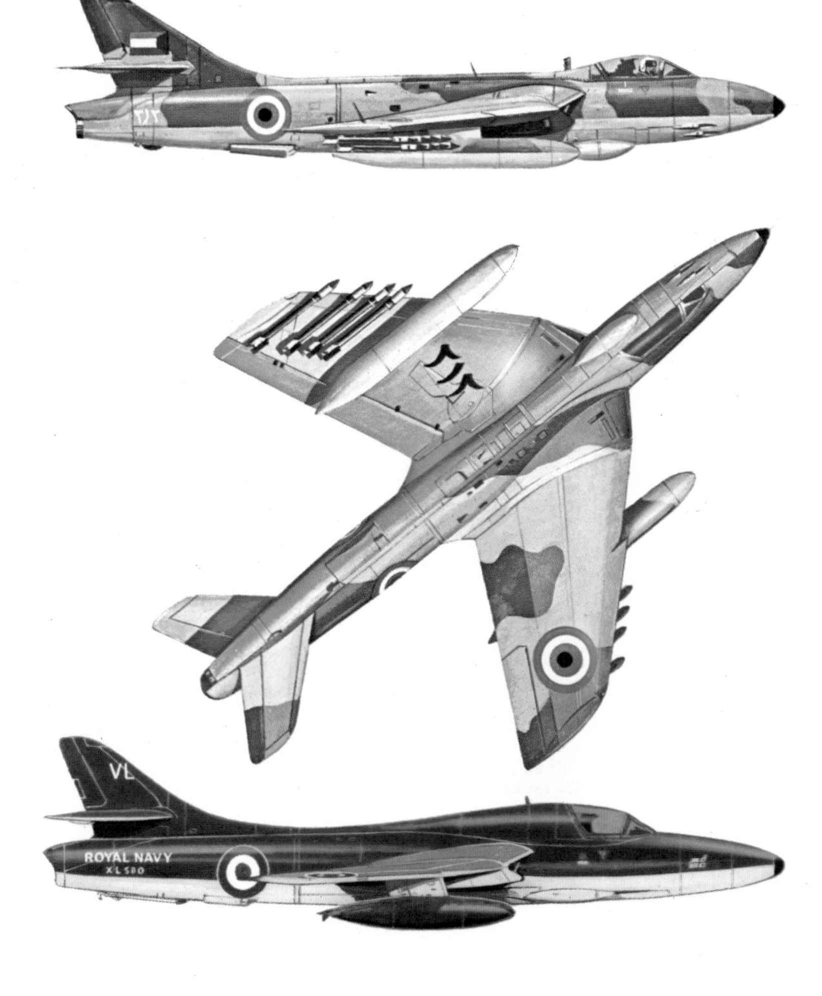

Top elevation and plan
Hawker Hunter MK 57 of the Kuwait Air Force, *circa* 1965
Lower elevation
Hawker Hunter T Mk 8, personal aircraft of the Flag Officer Naval Flying Training, Royal Navy, *circa* 1965

Span: 33ft 8in (10.26m)
Length: Mk 57, 45ft 10½in (13.98m); T Mk 8, 48ft 10½in (14.9m)
Weight, normal take-off: T.Mk 8, 17,200lbs (7,802kg)
Weight: Mk 57, 24,000lb (10,886kg)
Engine: Mk 57, 10,000lb (4,536kg) st Rolls-Royce Avon Mk 203 turbojet; T Mk 8, 7,550lb (3,425kg) Rolls-Royce Avon Mk 122 Turbojet
Max speed: 710mph (1,143kmh)
Operational ceiling: Mk 57, 51,000ft (15,695m); T Mk 8, 47,000ft (14,325m)
Range with external tanks: Mk 57, 1,840 miles (2,960km); T Mk 8, approx 1,650 miles (2,655km)
Armament: Mk 57, 4×30mm Aden cannon; 24×3in RPs under outer wings; T Mk 8, 1×30mm Aden cannon
Max bomb load: Mk 57, 2,000lb (907kg); T Mk 8, none

The Hunter deserves the term 'thoroughbred'. It was an outstanding export and military success for many years. The first of three Hawker P1067 prototypes flew on 20 July 1951, and the first production Mk 1 on 16 May 1953. Early production Hunters were specifically transonic day fighters; the Hawker-built Mks 1 (139 built) and 4 (365) had Avons, while the 45 Mk 2s and 105 Mk 5s built by Armstrong Whitworth had Sapphires. The Mk 4 was produced for the RAF, Denmark (30 Mk 51), Peru (16 Mk 52) and Sweden (120 Mk 50). A more powerful Avon, a 'dog-tooth' wing leading edge and a 'flying' tail marked the F Mk 6, the major single-seat version, which could carry a heavier range of external stores. The prototype flew on 22 January 1954; 264 were built by Hawker, 119 by Armstrong Whitworth. The F(GA) Mk 9 was a ground-attack conversion for the RAF in the Middle and Far East; the FR Mk 10 a three-camera reconnaissance conversion of the F Mk 6; and the GA Mk 11 an FAA attack trainer, converted from the F Mk 4 with no cannon but extensive stores. Production, including 2-seaters and 445 Mks 4 and 6 built in Holland and Belgium, totalled 1,977. Equivalents of the F Mk 6 were supplied to India (160 Mk 56), Iraq (35 Mk 59), Jordan (2 Mk 60), Lebanon (13 Mk 59) and Switzerland (100 Mk 58). India and Switzerland later re-ordered. Versions of the F (GA) Mk 9 and FR Mk 10 included four Mk 57s for Kuwait, and others for Abu Dhabi (Mk 76 and Mk 76A), Chile (Mks 71/71A), Jordan (Mks 73/73A/73B), Qatar (4 Mk 78), Rhodesia (12) and Singapore (16 Mks 74/74A). Many of those exported were diversions from RAF orders or refurbished ex-RAF aircraft. Hunters entered RAF service in July 1954, enjoying a length of service equalled by few aircraft. The Hunter is extremely versatile in the attack role, carrying a considerable variety of stores on four underwing stations, typically 16 60lb rockets, two 500 or 1,000lb bombs, napalm tanks or auxiliary fuel tanks. Some Dutch and Swedish F Mk 4s carried Sidewinder AAMs. A 2-seat Hunter was a logical development. The prototype T Mk 7 flew in 1955, an F Mk 4, with a longer front fuselage for a dual-control side-by-side cockpit and one Aden. Fifty entered RAF service from mid-1958; 20 went to the Dutch; the FAA counterpart was the 'hooked' T Mk 8 (28 built); Denmark had two Mk 53s, and Peru one Mk 62. Similar modifications to the F Mk 6 yielded the Mk 66 (India), Mk 66B (Jordan) Mk 66C (Lebanon), Mk 66D (India), Mk 67 (Kuwait), Mk 69 (Iraq/Lebanon), Mk 70 (Jordan), Mk 72 (Chile), Mk 75 (Singapore), Mk 77 (Abu Dhabi) and Mk 79 (Qatar). Total production of 2-seaters including conversions, was 101.

North American F-100 Super Sabre

North American F-100D Super Sabre of the 308th Tactical FS, 31st Tactical TW, USAF, Vietnam, 1969-70

Span: 38ft 9⅜in (11.82m)
Length (excluding probe): 49ft 4in (15.04m)
Weight, normal take-off: 34,050lb (15,445kg)
Engine: 11,700/16,950lb (5,307/7,688kg) st Pratt & Whitney J57-P-21A afterburning turbojet
Max speed: 910mph (1,464kmh) at 35,000ft (10,670m)
Operational ceiling: 36,000ft (11,000m)
Radius on internal fuel: 534 miles (860km)
Armament: 4×20mm M39E cannon
Max bomb load: 6×1,000lb bombs, or 2×Sidewinder or Bullpup AAMs or rocket pods

The Super Sabre, whose YF-100A prototype (the first of two) flew on 25 May 1953 after four years design work, deserves its niche in aviation history as the first fully supersonic warplane to enter quantity production. It began life under the company designation Sabre 45, the figure indicating the degree of wing sweepback, but subsequent development rendered it a completely new design. First production model was the F-100A day fighter (203 built) with 9,700lb (4,400kg) st J57-P-7 or -39 engines, four 20mm M-39E cannon and six stores attachment points. The F-100A, first flown on 29 October 1953, entered USAF service in September 1954; 80, brought up to F-100D standard, were supplied to the Chinese Nationalist Air Force. The F-100C, first flown on 17 January 1955, was a fighter-bomber version with a 7,500lb (3,402kg) external load on eight strong-points and able to refuel in flight. Production totalled 476, of which 260 were later released to the Turkish Air Force; they latterly served with the ANG. A number of design refinements, including a taller fin, appeared in the F-100D (first flight 24 January 1956), which entered production in 1956. In addition to the standard quartet of 20mm cannon, the F-100D could carry four Sidewinders and, for attack missions, two Bullpup missiles and/or a wide variety of conventional weapons up to a maximum of 7,500lb (3,402kg). The F-100D was the major production version, 1,274 being built, and was used extensively by the USAF in Vietnam in 1966-71, where the F-100 flew more sorties than had the North American P-51 in World War 2, by day and night. Others were supplied under MAP terms to the French and Danish air forces. Final version of the Super Sabre, which brought total production to 2,292 before completion in October 1959, was the 2-seat F-100F, which flew for the first time on 7 March 1957 following the flight of a TF-100C prototype on 6 August 1956: 339 F-100Fs were built. With two 20mm cannon and a 6,000lb weapons load, this model could perform either tactical attack roles or combat training duties and served in Vietnam in ground attack, fighter, electronic warfare and forward air control roles. The Danish and Turkish air forces, as well as the USAF, received quantities of the F-100F.

Tupolev Tu-16 (Badger)

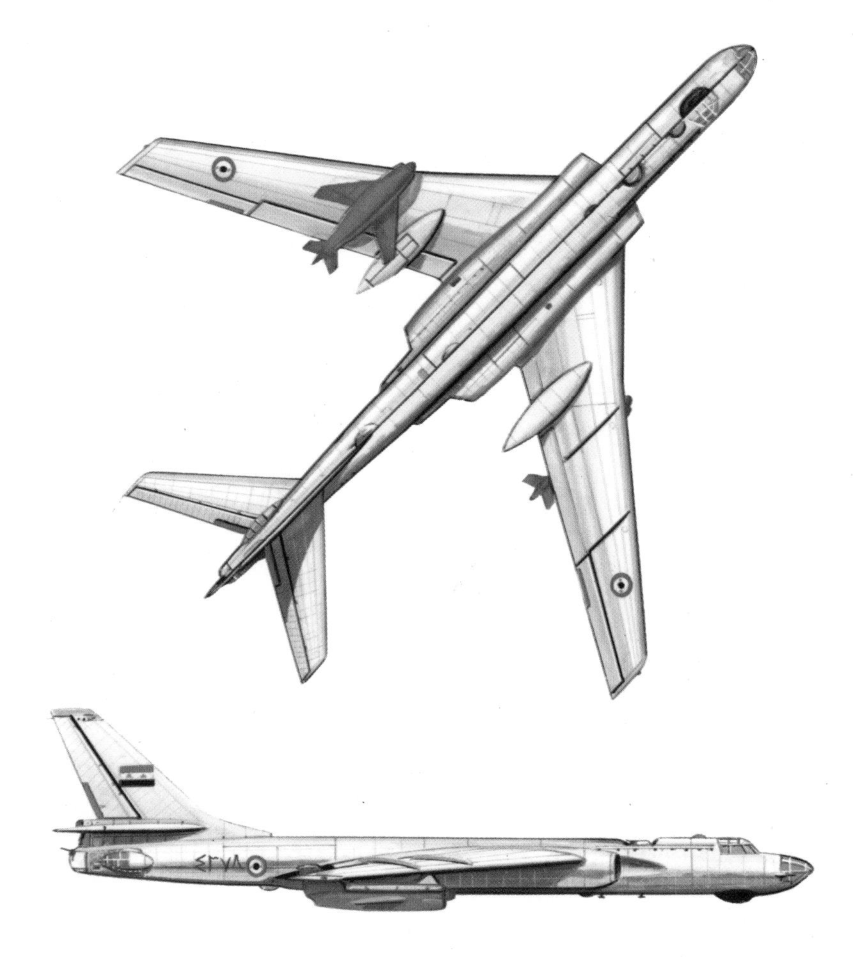

Tupolev Tu-16 (Badger-A) of the Egyptian Air Force, *circa* 1967

Span: 108ft 0½in (32.93m)
Length: 114ft 2in (34.80m)
Weight, normal take-off: 158,733 (72,000kg)
Engines: 2×19,290lb (8,750kg) st Mikulin AM-3M turbojets
Max speed: 616mph (992kmh) at 19,685ft (6,000m)
Operational ceiling: 40,355ft (12,300m)
Range with max bomb load: 2,985 miles (4,800km)
Armament: 6 or 7×23mm NR-23 cannon
Max bomb load: 19,800lb (9,000kg) internally; or underwing or under-fuselage ASMs

Produced originally for the V-VS's *Dalnaya Aviatsiya* (Long Range Aviation), about 2,000 Tu-16s are believed to have been built in the USSR. Replacing the Tu-4, the Tu-16 (design bureau designation Tu-88) entered Soviet Air Force service in 1954-55, the first production version being code-named Badger-A. The first production Soviet all-swept bomber, it was used as a bomber and missile carrier. Carrying a six or seven-man crew, the Badger-A had twin 23mm cannon in dorsal, central and tail positions, with a seventh gun in the starboard nose; a 21ft (6.40m) weapons bay accommodated a maximum 19,840lb (9,000kg) war load. About 20 Badger-As, supplied to Egypt, were destroyed in the Six-Day War of 1967, but were subsequently replaced; six others were supplied to Iraq. A number of Soviet Air Force machines were converted for the reconnaissance role with additional radar and long-range fuel tanks; all versions of the Tu-16 have wingtip-to-wingtip in-flight refuelling capability. The Tu-16 began a new lease of life in the early 1960s as an anti-shipping strike and long-range maritime reconnaissance aircraft with the A-VMF (Soviet Naval Aviation), and is numerically A-VMF's most important aircraft. Badger-B is a naval conversion carrying two Kennel anti-shipping missiles on pylons outboard of the main undercarriage fairings; it also equipped two squadrons of the Indonesian Air Force from 1961. Badger-C is a more extensive conversion, featuring a large search radar in a longer, solid nose, and is a carrier for the Kipper stand-off anti-shipping missile, one being attached beneath the centre fuselage. Pairs of Badgers have frequently been observed on long-range naval patrol, fitted with auxiliary fuel tanks and fairings which contain ECM equipment; they are used to test NATO borders and western defences. ECM reconnaissance models include Badger-D and -E, similar to -C but with various under-fuselage fairings, and -F, which carries ECM pods on wing-mounted pylons but which is otherwise similar to Badger-A. Badger-G is a successor to Badger-B carrying two Kelt rocket-powered anti-shipping weapons on larger underwing pylons. The A-VMF had in 1974 about 275 of this version, and others were supplied to Egypt. About 60 Tu-16s have been built since 1968 (as the Harbin B-6) in the Chinese People's Republic, which in 1973 supplied about a dozen to the Pakistan Air Force.

MiG-19 (Farmer)

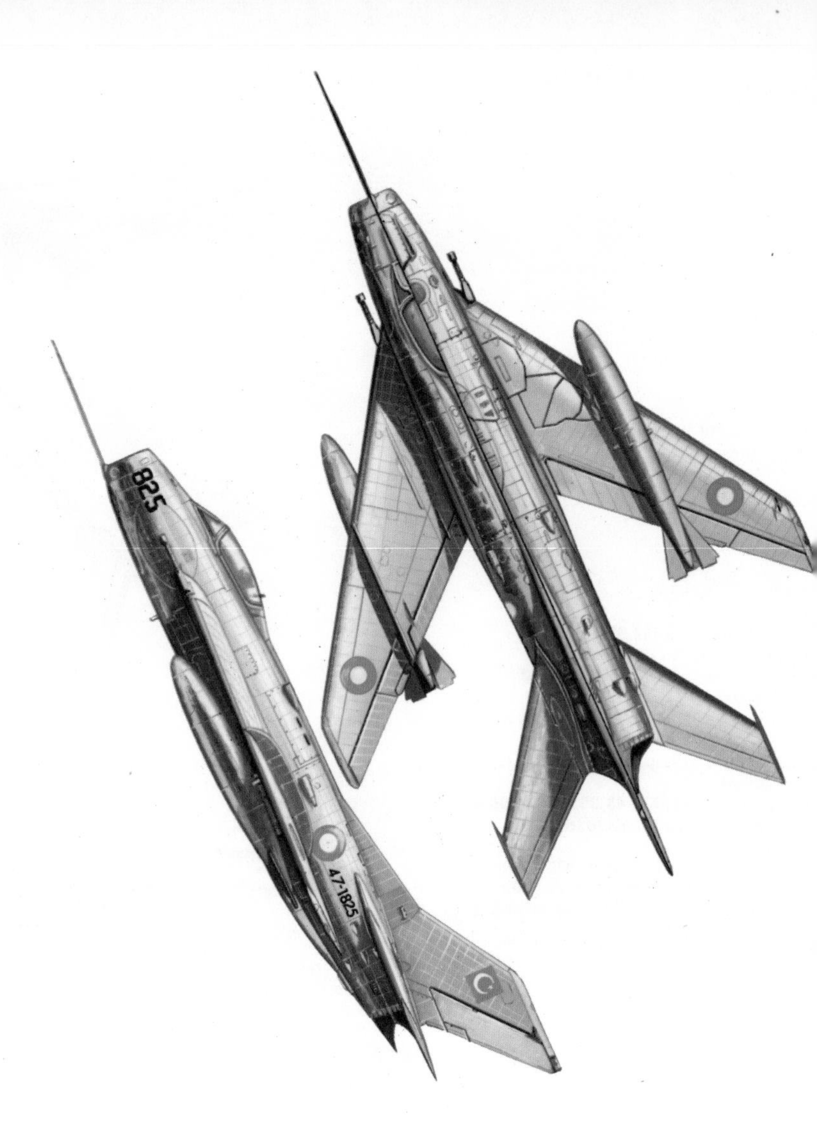

Shenyang F-6 (Chinese-built Mikoyan MiG-19SF Farmer-C) of the Pakistan Air Force, 1968

Span: 30ft 2¼in (9.20m)
Length (excluding nose probe): 41ft 4in (12.60m)
Weight: 19,180lb (8,700kg)
Engines: 2×5,732/7,165lb (2,600/3,250kg) st Tumansky R-9B afterburning turbojets
Max speed: 902mph (1,452kmh) at 32,800ft (10,000m)
Operational ceiling: 58,725ft (17,900m)
Combat radius with 2 underwing tanks: 425 miles (685km)
Armament: 2 or 3×30mm NR-30 cannon (plus 2×Sidewinder AAMs on Pakistan aircraft)
Max bomb load: 2×250kg or 500lb bombs, 2×212mm rockets, or 16 smaller rockets

In 1950 the Mikoyan design bureau began work on the first Soviet supersonic fighter to achieve squadron status. The design, using two small-diameter Mikulin AM-5 turbojets side by side in the fuselage, received official approval on 30 July 1951; the first prototype, designated I-350, flew in September 1953. A small initial production batch, with 4,850/6,700lb (2,200/3,040kg) st AM-5Fs, entered service in early 1955, first being seen by Western observers at the Aviation Day display that summer. This model suffered from elevator troubles, and with its unbalanced armament of one 37mm and two 23mm cannon was apparently not produced in great numbers. A new model, the MiG-19S (Farmer-A) day fighter, entered service in 1956. This retained the AM-5s but had three of the faster-firing 30mm NR-30 cannon, a one-piece 'slab' tailplane, wing spoilers and a third, ventral airbrake supplementing those on the fuselage aft of the wings. Two years later this was succeeded by the MiG-19SF (Farmer-C), which became the standard day fighter version, the designation indicating more powerful RD-9Bs. Both the MiG-19S and SF had two stores points beneath each wing, the outer pair more usually carrying fuel tanks and the inboard pair, unguided air-to-air rockets. Contemporary with the SF was a limited all-weather version, the PF (Farmer-B), with a nose section some 1ft 9½in (0.55m) longer, a 'lipped' air intake and a small, hemispherical centrebody radome mounted on the air intake splitter plate. The nose gun was deleted, and two leading-edge weapon-launching shoes, usually carrying unguided rockets pods, replaced the inboard pylons. A more fully-developed all-weather version was the MiG-19PM (Farmer-D), in which the two wing-root guns were also deleted and the number of inboard leading-edge pylons increased to four, each capable of carrying an Alkali AAM. MiG-19 production was on a large scale; believed to have ended in the USSR in 1959, it included licence production in Czechoslovakia as the S-105. Several versions are still in production in China, as the Shenyang F-6 and TF-6. The Nanchang A-5 (Fantan) attack fighter is an extensively redesigned F-6. About 90 F-6s equivalent to the MiG-19SF were supplied to Pakistan (with provision for Sidewinder AAMs and including ground-attack and fighter-reconnaissance models), and Tanzania (about 30). Soviet-built MiG-19s were supplied to Bulgaria, Cuba, Egypt, the German Democratic Republic, Indonesia, Iraq, North Korea, Romania and Yugoslavia; it was replaced in Soviet service by the MiG-21.

Boeing B-52 Stratofortress

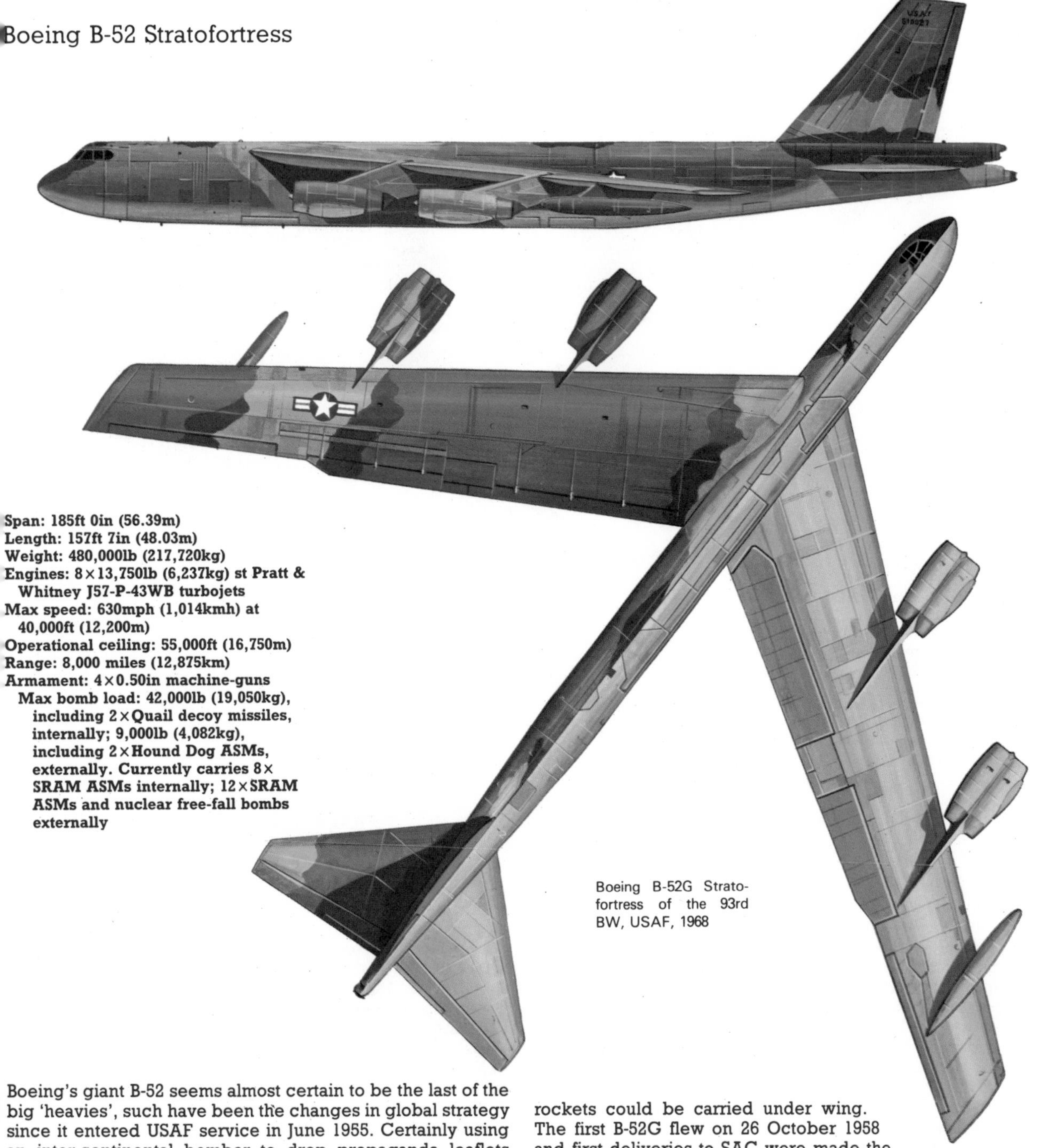

Span: 185ft 0in (56.39m)
Length: 157ft 7in (48.03m)
Weight: 480,000lb (217,720kg)
Engines: 8×13,750lb (6,237kg) st Pratt & Whitney J57-P-43WB turbojets
Max speed: 630mph (1,014kmh) at 40,000ft (12,200m)
Operational ceiling: 55,000ft (16,750m)
Range: 8,000 miles (12,875km)
Armament: 4×0.50in machine-guns
Max bomb load: 42,000lb (19,050kg), including 2×Quail decoy missiles, internally; 9,000lb (4,082kg), including 2×Hound Dog ASMs, externally. Currently carries 8× SRAM ASMs internally; 12×SRAM ASMs and nuclear free-fall bombs externally

Boeing B-52G Stratofortress of the 93rd BW, USAF, 1968

Boeing's giant B-52 seems almost certain to be the last of the big 'heavies', such have been the changes in global strategy since it entered USAF service in June 1955. Certainly using an inter-continental bomber to drop propaganda leaflets over North Vietnam was the antitheses of cost-effectiveness, and far from the B-52's designed role. The XB-52 prototype flew on 2 October 1952, followed by one YB-52, three B-52As and 50 B-52B and RB-52B, initial production models. SAC had in 1974 about 450 B-52s, approximately a third being the tall-finned B-52C, D, E and F (respective production totals 35, 170, 100 and 89), differing in power, electronics and equipment; their maximum internal bomb load is 60,000lb (27,216kg). The remainder were B-52Gs or Hs; the former has a marked performance increase and a cropped vertical tail, and was the first model to carry the Hound Dog stand-off missile. The G's internal load included two Quail decoy missiles and 20,000lb (9,072kg) of bombs and has a tail turret with four remotely-controlled 0.50in machine-guns. Decoy rockets could be carried under wing.
The first B-52G flew on 26 October 1958 and first deliveries to SAC were made the following February; 193 were built. The final version, the B-52H (first flight 6 March 1961; 102 built up to June 1962) is similar to the G except for 16,000lb (7,257kg) st Pratt & Whitney TF33-P-1 turbofans, further increasing range and performance, and a multi-barrel cannon in the tail. The B-52H has flown more than 12,500 miles (20,117km) unrefuelled. All serving B-52s have been modernised and strengthened for low-level penetration missions. A modernisation programme on 96 G and H models enables carriage of up to 20 Short-Range Attack Missiles. Some 200 G and H models have Hughes FLIR (foward-looking infra-red) sensors in twin under-nose fairings and EVS (Electro-optical Viewing System) night vision equipment installed. In 1975-77 80 B-52Ds were rebuilt to extend their lives.

McDonnell Douglas A-4 Skyhawk

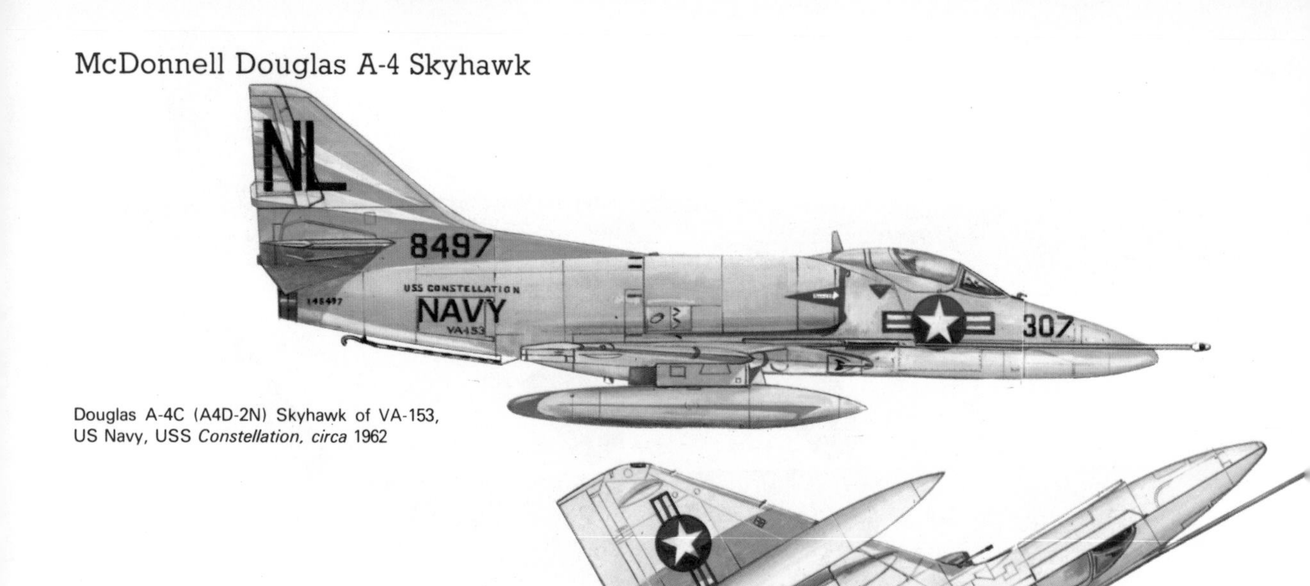

Douglas A-4C (A4D-2N) Skyhawk of VA-153, US Navy, USS *Constellation, circa* 1962

Span: 27ft 6in (8.38m)
Length (including probe): 42ft 10¾in (13.07m)
Weight, normal take-off: 17,295lb (7,845kg)
Engine: 7,700lb (3,493kg) st Wright J65-W-164 turbojet
Max speed: 680mph (1,094kmh) at sea level
Range: 1,150 miles (1,850km)
Armament: 2×20mm Mk 12 cannon
Max bomb load: 5,000lb (2,268kg) externally

The Skyhawk, jet successor to the Douglas Skyraider attack bomber, was an extremely effective exercise by designer Ed Heinemann in weight saving and compact design, small enough to fit into USN carrier lifts without wing folding, yet powerful and with great load carrying capability. The USN placed a contract for one prototype XA4D-1 in June 1952; this first flew on 22 June 1954 (the only prototype). The A4D-1 (165 built, including 19 YA4D-1s) initial production model with 7,700lb (3,500kg) st Wright J65-W-4 or 4B, entered USN service in October 1956. The A4D-2 (542 built) had an in-flight refuelling probe. The A4D-2N (638 built; first flown 21 August 1959) was a limited all-weather version with radar equipment in a slightly longer nose. All had J65 turbojets, two 20mm cannon in the wing roots, and one stores point under the fuselage and two underwing. The next major production model was the A4D-5 (496 built), introducing important changes, chiefly the 8,500lb (3,855kg) st Pratt & Whitney J52-P-6A, increasing range 25 per cent, and four stores pylons underwing; the first flew on 12 July 1961. When the USN and USAF adopted a unified designation system in 1962, the A4D-1, -2, -2N and -5 became the A-4A, B, C and E, A-4D being avoided to prevent confusion. Early in 1964 the USN ordered two TA-4Es, a tandem 2-seat operational trainer with lengthened fuselage. Deliveries of the production version, the TA-4F, began in 1966; a simplified trainer, the TA-4J, flew in May 1969. The A-4F (146 built), flown on 31 August 1966, was an updated single-seat attack model, with operational and avionics improvements including a 'saddle-back' dorsal avionics fairing, wing-spoilers, a zero-zero ejection seat and pilot armour. Several previous marks were updated to A-4F standard. The A-4M Skyhawk II for the USMC has an 11,200lb (5,080kg) st J52, deeper canopy, square-top fin, brake parachute, increased cannon ammunition, five stores points, and other refinements. Deliveries began in November 1971. Gaps in the sequence are filled by converted and/or exported versions. The Argentine Air Force and Navy, respectively, received 75 and 16 ex-USN A-4Bs, redesignated A-4P and Q respectively. The RAN received 16 A-4Gs and four TA-4Gs, half ex-USN, half new. The A-4H for Israel (90, plus 10 TA-4H), is based on the A-4E but has a brake parachute, square-top fin, and 30mm DEFA cannon. Israel also acquired about 30 A-4Es, some converted later to A-4F standard. Ten A-4Ks and four TA-4Ks, corresponding to the F, went to the RNZAF in 1970; Kuwait received 30 A-4KU and six TA-4KU. One hundred A-4Cs brought up to A-4K standard for the USN Reserve are designated A-4L. The A-4N Skyhawk II, flown on 8 June 1972, is an A-4M with twin 30mm DEFA cannon, and an updated navigation and weapons delivery system; Israel ordered 117. The A-4S is a refurbished A-4B; Singapore ordered 40. First in action during the 1958 Lebanon crisis, the Skyhawk was extensively used in Vietnam for ground-attack and support. Extended many times, production ceased in 1979.

Hawker Siddeley Vulcan

Hawker Siddeley Vulcan B Mk 2 of No 35 Sqdn, RAF, Tengah, Singapore, 1967

Span: 111ft 0in (33.83m)
Length: 99ft 11in (30.45m)
Weight: approx. 190,000lb (86,180kg)
Engines: 4×20,000lb (9,072kg) st Rolls-Royce (Bristol) Olympus Mk 301 turbojets
Max speed: 645mph (1,038kmh)
Operational ceiling: 65,000ft (19,800m)
Low-level radius on internal fuel: 1,725 miles (2,775km)
Armament: None
Max bomb load: 21,000lb (9,525kg) internally

The transonic Vulcan is the world's largest operational delta-winged aeroplane, and entered Bomber Command service in the middle 1950s. Designed by the former Avro company, the Avro 698 first prototype (VX770) made its maiden flight on 30 August 1952 powered by four Rolls-Royce Avon engines. The second prototype (first flight 4 September 1953) and subsequent aircraft all had Bristol Siddeley Olympus turbojets. Production of the B Mk 1 began in 1953; 45 were built (XA889-913, XH475-483, XH497-506 and XH532), the first operational unit being No 83 Sqdn in May 1957. Most B Mk 1s were subsequently modified to B Mk 1A standard, with electronic countermeasures equipment in a bulged tail cone. By the mid 1960s all Mk 1/1A Vulcans had been placed on the reserve. Their place in the front-line strategic V-bomber force was then occupied by the B Mk 2, which has a 12ft (3.66m) greater wing spread and more powerful Olympus engines. Deliveries of the B Mk 2, which began in July 1960, were made to Nos 9, 12, 27, 35, 44, 50, 83, 101 and 617 Sqdns; 79 were built (XH533-539, XH554-563, XJ780-784, XJ823-825, XL317-321, XL359-361, XL384-392, XL425-427, XL443-446, XM569-576, XM594-612 and XM645-657). Whereas the Mks 1/1A were able to carry free-falling nuclear weapons or 21,000lb (9,525kg) of conventional high explosive bombs, the B Mk 2 was equipped also to carry the Blue Steel stand-off bomb. Both marks were, while in service, refitted with successively more powerful variants of the Olympus jet engine, and the Mk 2 with capacity for in-flight refuelling. In 1971 RAF Strike Command still retained two Blue Steel-equipped Vulcan squadrons, but by 1974 all remaining Vulcan bomber squadrons (Nos 9, 35, 44, 50, 101 and 617) were fulfilling a low-level overland strike role with conventional or tactical nuclear weapons rather than a strategic bombing role. A seventh squadron, No 27, operated in the strategic reconnaissance role with SR Mk 2 aircraft, converted from B Mk 2s. All Vulcans, along with Buccaneers, are scheduled to be replaced by the Panarvia Tornado in the mid-1980s.

Lockheed F-104 Starfighter

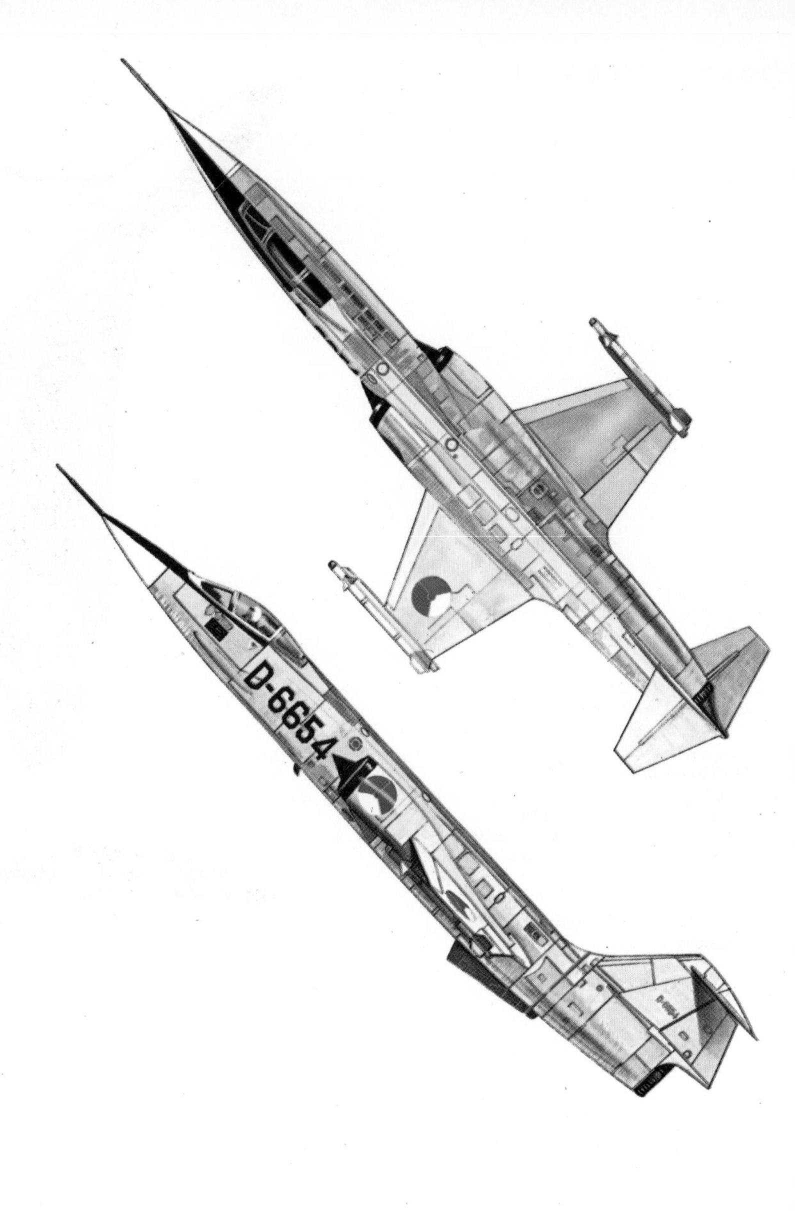

Fiat-built Lockheed F-104G Starfighter of No 323 Sqdn, Royal Netherlands Air Force, 1964

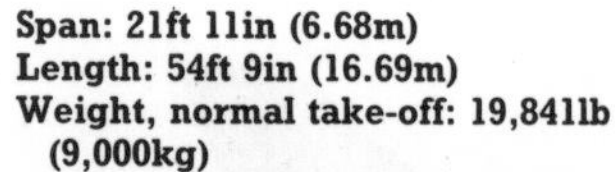

Span: 21ft 11in (6.68m)
Length: 54ft 9in (16.69m)
Weight, normal take-off: 19,841lb (9,000kg)
Engine: 10,000/15,800lb (4,536/7,167kg) st General Electric J79-GE-11A afterburning turbojet
Max speed: 1,320mph (2,125kmh) above 36,000ft (11,000m)
Operational ceiling: 55,000ft (16,765m)
Typical operational radius: 690 miles (1,110km)
Armament: 1×20mm M61 Vulcan multi-band cannon; position for 2 or 4 AAMs
Max bomb load: 4,000lb (1,814kg) externally

Lockheed's Starfighter is an instance of an aeroplane that disappointed in its intended role only to excel later at a quite different task. Thus only 277 of the first four marks were built, as short-range day interceptors and fighter-trainers, but the multi-mission F-104G, for tactical support and reconnaissance missions, became the subject of the biggest international co-operative manufacturing programme in Europe since World War 2. Lockheed built two XF-104s with 10,000lb (4,409kg) st Wright XJ65-W-6s, the first flying on 7 February 1954, followed by 15 YF-104s for evaluation. Initial production models, for Air Defense Command and TAC of the USAF, were the single-seat F-104A (153 built, first flight 17 February 1956); 2-seat F-104B (26 built, first flight 7 February 1957); single-seat F-104C (77 built); and 2-seat F-104D (21 built). Jordan bought 36 As, Pakistan 12 As, and Taiwan 25 Bs. Mitsubishi assembled 20 F-104DJs for the JASDF from Lockheed-built components; and 30 F-104Fs, similar to the D, were built by Lockheed for the German *Luftwaffe*. Major changes introduced with the single-seat F-104G were a strengthened airframe, upward (instead of downward) ejection seat, and enlarged vertical tail. The G was built, by German, Dutch and Belgian consortia. for the *Luftwaffe* (604), Holland (95) and Belgium (99); and for Italy (124) and the Netherlands (25) by Fiat and Aermacchi in Italy. Lockheed-built Gs were supplied to Germany (96), Belgium (one) and Italy (one). Lockheed also built 181 2-seat TF-104Gs. F-104G variants included the CF-104 (200 built by Canadair for the RCAF, which also received 38 Lockheed-built CF-104D trainers); and the F-104J for the JASDF, which has 210, mostly licence-built by Mitsubishi. Canadair production included US-financed Gs for Denmark (25), Greece (36), Norway (16), Spain (25) and Turkey (38). Standard fixed armament of the F-104G (first flight 5 October 1960) is a 20mm Vulcan multi-barrel cannon, augmented by two or four Sidewinders for interception. In the attack role two Bullpup missiles, two rocket pods, three 1,000lb (or two 1,000lb and one 2,000lb) bombs, three land mines or one tactical nuclear weapon are among the many external loads. Production of all these versions is complete, but Aeritalia built F-104Ss for the AMI (205) and Turkey (40). This is a strengthened, all-weather interceptor with a 17,900lb (8,120kg) st J79-GE-19 or J1Q and nine external armament points; the first flew on 30 December 1968. Weapons include bombs and rockets, and Sidewinder or Sparrow AAMs.

Convair F-106 Delta Dart

General Dynamics (Convair) F-106A Delta Dart of the 94th Fighter Interceptor Sqdn, Air Defense Command, USAF, *circa* 1962

Span: 38ft 3½in (11.67m)
Length (including nose probe): 70ft 8¾in (21.56m)
Weight, normal take-off: 35,500lb (16,103kg)
Engine: 17,200/24,500lb (7,802/11,113kg) st Pratt & Whitney J75-P-17 afterburning turbojets
Max speed: 1,525mph (2,455kmh)
Operational ceiling: 57,000ft (17,375m)
Combat radius, standard fuel: 575 miles (925km)
Armament: No guns; 4×Falcon AAMs and 2×Genie rockets internally

Close family resemblance between the F-102 and F-106 is no coincidence, for the F-106 began its development in 1955 as the F-102B. Such were the ultimate differences between the two, however, that a new F designation was allotted. Although retaining the delta-wing and area-ruled fuselage configuration of the subsonic F-102, the F-106's fuselage was entirely redesigned to take the Pratt & Whitney J75, delivering 50 per cent more power than the F-102's J57. To optimise performance at all speeds, the engine intakes were closer to the engine and given variable ducts. The greater power and increased aerodynamic efficiency resulted in the F-106 being almost twice as fast as the F-102. In December 1957 an F-106A set a world absolute speed record of 1,525mph (2,454kmh). It is still one of the most advanced interceptors. Its electronic guidance and fire control equipment, designed to work with the SAGE (Semi-Automatic Ground Environment) defence system, detects its target by radar and launches its four Falcon and/or two Genie missiles entirely automatically. An entire combat mission, between take-off and touchdown, can be flown without the pilot touching a single control. The first to fly, on 26 December 1956, was a production F-106A. The F-106B (first flight 9 April 1958) tandem 2-seat combat trainer has the same combat capability as the A, and is almost as fast. When production ended late in 1960 277 single-seat F-106As and 63 F-106Bs had been built. First deliveries were made in July 1959, and by mid-1961 Delta Darts equipped about half the all-weather squadrons of the USAF's Air Defense Command. The F-106A has undergone two major rebuild programmes, giving updated avionics, new digital air data links, improved navigation/identification/ECM/communications systems, new IR sensors, new missiles, the installation of zero-height crew escape systems and provision for underwing drop-tanks and in-flight refuelling capability. In 1970 tests were conducted with an F-106A (58-795) fitted with an improved visibility cockpit hood, optical gun-sight and an M-61A 20mm multi-barrel cannon in a semi-retractable ventral fairing. This gun installation was adopted as a standard fit for in-service aircraft in 1973. In the late 1970s the F-106A/B equipped all but one of Aerospace Defense Command's seven Fighter Interceptor Squadrons, all based in the USA. In 1980 ADC was merged with TAC. Additionally, five ANG squadrons have flown F-106s.

MiG-21 (Fishbed/Mongol)

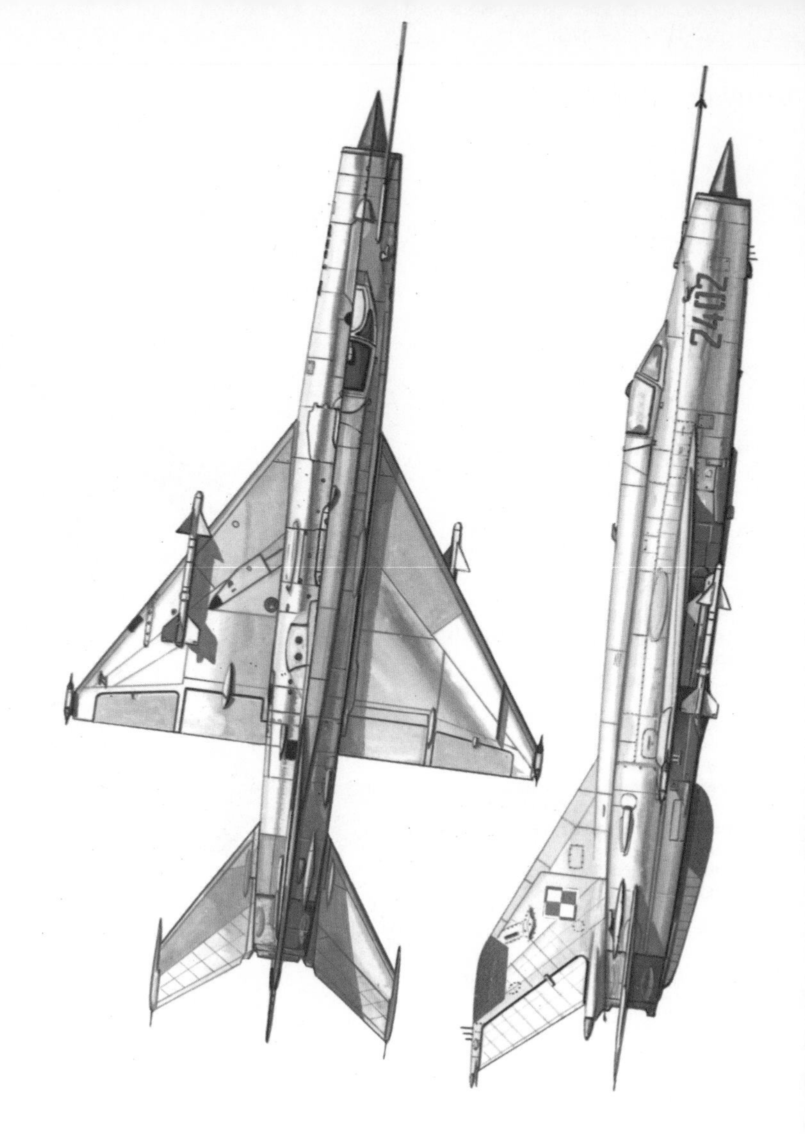

Mikoyan MiG-21MF (Fishbed-K) of the Polish Air Force, 1973

Span: 23ft 5½in (7.15m)
Length (excluding nose probe): 44ft 2in (13.46m)
Weight: 20,725lb (9,400kg)
Engine: 11,244/14,550lb (5,100/6,600kg) st Tumansky R-13-300 afterburning turbojet
Max speed: 1,385mph (2,230kmh) above 36,000ft (11,000m)
Operational ceiling: approx 50,000ft (15,240m)
Combat radius: 350 miles (560km)
Armament: 1×twin-barrel 23mm GSh-23 cannon; 2×Atoll and 2×Advanced Atoll (or 4×Advanced Atoll) AAMs Max bomb load: 2×500kg and 2×250kg bombs; 4×ASMs; or 4×16-rocket packs

One of the most extensively-built and widely-used combat aircraft, the MiG-21 serves with the Soviet Air Force and over a score of associate countries. First flown in 1955, two differing E-5 prototypes (Fishbed-A and -B), were first seen by western observers at the June 1956 Aviation Day display at Tushino. The first standard production model was the MiG-21F (Fishbed-C), the suffix indicating a more powerful engine. The MiG-21F was supplied to Afghanistan, Algeria, the Chinese People's Republic, Cuba, Czechoslovakia, Egypt, Finland, the German Democratic Republic, Hungary, India, Indonesia, Iraq, North Korea, Poland, Romania, Syria and Yugoslavia and was built in China and Czechoslovakia. The MiG-21F is a short-range clear-weather day interceptor, armed with either one or two 30mm NR-30 cannon in the nose and a K-13 Atoll infra-red homing missile or a pod of 16 57mm rockets beneath each wing. There is a third ventral stores attachment point, usually occupied by an auxiliary fuel tank to offset the modest internal fuel load. A tandem 2-seat trainer version is designated MiG-21UTI (Mongol-A). The improved version designated MiG-21PF (Fishbed-D) followed into service, the P suffix letter denoting an all-weather F version. It has a more powerful RD-11, and is distinguishable by its less-tapered nose, larger-diameter air intake and much larger centre-body radar, broader-chord fin (on most aircraft) and a fairing aft of the cockpit. No fixed gun armament is fitted, but pods of 16 57mm rockets may be alternated with Atoll missiles on the underwing pylons. Fishbed-Ds were supplied to Bulgaria, Egypt, East Germany, Hungary, Poland and North Vietnam. Further development yielded Fishbed-F (MiG-21PFM), H (MiG-21R and RF), J (MiG-21PFMA and MiG-21MF), and K. A Czech-built Fishbed-F version designated MiG-21SPS, has blown flaps. Fishbed-H is a tactical reconnaissance version of Fishbed-J. Fishbed-K is similar to the J but carries wingtip ECM pods and has a longer dorsal spine. The Fishbed-J has a strengthened wing giving supersonic capability at sea level and incorporates improved features and has a deeper, straight-topped dorsal spine fairing. It has a multi-mission capability, underwing attachments being increased to four, carrying interception and/or ground attack weapons. Late production models have a 23mm twin-barrel GSh-23 cannon in a ventral fairing. The MiG-21MF Fishbed-K has a constant-depth dorsal spine and more powerful RD-13. Two-seat variants of the broad-finned MiG-21s are code-named Mongol-B. Chinese production (un-licensed) began with the Fishbed-C (Shenyang F-7). HAL in India imported ten MiG-21Fs and two PFs in 1963-64, before licence-building 196 PFs, following in 1973 with the PFMA and M versions.

BAC (English Electric) Lightning

BAC (English Electric) Lightning F Mk 6 of No 74 Sqdn, RAF, *circa* 1968

Span: 34ft 10in (10.61m)
Length (including probe): 55ft 3in (16.84m)
Weight: approx 48,000lb (21,770kg)
Engines: 2×12,690/16,360lb (5,756/7,420kg) st Rolls-Royce Avon Mk 301 afterburning turbojets
Max speed: approx 1,450mph (2,335kmh)
Operational ceiling: over 60,000ft (18,300m)
Range with ventral tank: approx. 750 miles (1,200km)
Armament: 2×Firestreek or Red Top AAMs; 44×2in, or 36×68mm rockets, or 2× m/gun pods; provision for 2×30mm Aden cannon in ventral pack
Max bomb load: 2,000lb (907kg)

The twin-engined Lightning, the RAF's front-line fighter for more than a decade, was the first British-designed combat aircraft capable of Mach 2. Stemming from the English Electric P1A, which first flew on 4 August 1954, it owed much to research carried out, from 1952, with the Short SB5. Two Sapphire-engined P1As were built, followed by three P1B prototypes (first flight 4 April 1957) with Avons and 20 development aircraft. Initial production version was the F Mk 1 (20 built) first flown on 29 October 1959 and delivered to No 74 Sqdn, RAF Fighter Command from summer 1960. Armament was two 30mm Aden cannon and two Firestreak AAMs. The similar F Mk 1A (28 built), had in-flight refuelling. From the F Mk 1A were developed the F Mk 2 (14 built) and F Mk 2A (30 converted from F Mk 2), with longer range, improved reheat and avionics, the first example flying on 11 July 1961. These three models had round-topped fins. First model with the modified square-topped fin, of greater area, was the F Mk 3, which first flew in definitive production form on 16 July 1962. Powered by 16,360lb (7,420kg) st Avon 300-series, 63 F Mk 3s were built, entering service from April 1964. Armament of the Mks 2 and 3 was Hawker Siddeley Red Top missiles, and the guns were deleted. On 17 April 1964, the prototype F Mk 6 flew, a fully-developed version of the F Mk 3 of which 62 were completed with cambered, less-swept outer wings, a ventral fuel/weapons pack over double the capacity of that introduced on the F Mk 3, and provision for an arrester hook for short-field landings. The large overwing fuel tanks first tested on the F Mk 3 became standard on the F Mk 6, delivered from December 1965. RAF operational trainer versions, corresponding to the F Mks 1 and 3 respectively, designated T Mk 4 and T Mk 5, have side by side 2-seat cockpits with modified canopies in a wider front fuselage. There were two T Mk 4 prototypes, 20 production aircraft; a production T Mk 4 served as prototype for the 22 production T Mk 5s. Exports included five Mk 2s and two T Mk 4s supplied in 1966-67 as F Mk 52/T Mk 54 to Saudi Arabia, and the F Mk 53 and T Mk 55 special export models ordered by Saudi Arabia and Kuwait. First flights were made on 1 November and 3 November 1966 respectively by the F Mk 53 and T Mk 55; deliveries to Saudi Arabia (34 F Mk 53s and six F Mk 55s), began in December 1967. Kuwait ordered 12 F Mk 53s and two T Mk 55s. The F Mk 53 is a developed version of the F Mk 6, with Avon 302-Cs and multi-mission capability. By 1977 the Lightning had been replaced as the RAF's standard day fighter by the Phantom, but two squadrons remained. Production ended in 1972, totalling 388.

McDonnell Douglas F-4 Phantom

McDonnell Douglas (F-4K) Phantom FG Mk 1 of No 43 Sqdn, RAF, Leuchars, 1969

Span: 38ft 5in (11.70m)
Length: 57ft 11in (17.65m)
Weight: 58,000lb (26,308kg)
Engines: 2×12,500/21,250lb (5,670/9,638kg) st Rolls-Royce Spey RB 168-25R Mk 201 afterburning turbojets
Max speed: 1,386mph (2,230kmh) above 36,000ft (11,000m)
Operational ceiling: 71,000ft (21,640m)
Typical tactical radius: 550 miles (885km)
Armament: 4, 6 or 8×AAMs, depending on type (Sidewinder, Sparrow, Martel, etc)
Max bomb load: 16,000lb (7,257kg) of bombs, ASMs or rockets externally

McDonnell Douglas's second Phantom originated to a USN specification for a single-seat carrier-based attack fighter. In 1955 the USN altered this to a missile-armed fighter, changing the designation from AH-1 to F4H-1; 23 were ordered for tests, the first flying on 27 May 1958 with two J79-GE-3As. Dihedral on the outer wing, anhedral on the all-moving tailplane, a blown-flap system of boundary layer control and more powerful J79s were introduced. The trials batch, and the first 24 production Phantom IIs, became F4H-1Fs when fitted with J79-GE-2 or -2A turbojets; parallel versions were ordered by the USAF – adopting a production Navy fighter for the first time – as the F-110A and PR RF-110A. In 1962, the US services adopted a unified designation system, the F4H-1F becoming the F-4A, the definitive F4H-1 (with J79-GE-8s) the USN/USMC F-4B (635 built), the proposed F4H-1P 'camera job' the USMC's RF-4B, and the USAF versions F-4C and RF-4C. The USAF received 583 F-4Cs, for TAC, Pacific Air Force and USAFE; 36 were supplied to Spain. F-4C deliveries were completed in 1966. The USAF's F-4D, flown on 8 December 1965, had J79-GE-15s and improved radar and electronics; 825 were built, 64 being supplied to Iran and 18 to South Korea. The improved multi-role F-4E for the USAF, first flown on 30 June 1967, has more powerful J79-GE-17s, an M61 20mm multi-barrel cannon under the nose and extra fuel capacity. It was exported to Greece, Iran, Israel and Turkey, and Mitsubishi built 128 for the JASDF. The RF-4E, first flown in October 1970, is a multi-sensor reconnaissance version ordered originally by Federal Germany (88). Japan ordered 14 as the RF-4EJ, and others went to Iran. Israel ordered 168 F-4Es and RF-4Es. The F-4F for the *Luftwaffe* (175 built) has leading-edge slots. The F-4B has been updated. The USN's first updated model was the F-4G, which entered service in Vietnam in 1966. A more extensive update is the F-4J (J79-GE-10; flown 27 May 1966) for the USN and USMC. From 1973 178 F-4Bs were updated, designated F-4N. The F-4K and F-4M, based on the F-4B to British specifications with Rolls-Royce Spey turbofans, were ordered for the RAF/FAA. Deliveries of F-4Ks (FG Mk 1s) began in April 1968, but only one operational FAA squadron operated the type, the remainder being transferred to the RAF. The similar F-4M is designated FGR Mk 2 by the RAF; deliveries began in August 1968.

Dassault Mirage III and 5

Top elevation and plan
Dassault Mirage III-R of the *3ème Escadron, 33ème Escadre, Armée de l'Air,* Strasbourg, *circa* 1966-67
Lower elevation
Dassault Mirage 5-BA of the 2e Wing, Belgian Air Force, Floreenness, 1972

Span: 27ft 0in (8.22m)
Length (including probe): 49ft 3½in; III-R: 50ft 4¼in (15.5m); III-S 51ft 0¼in (15.55m)
Weight: 29,760lb (13,500kg)
Engine: 9,430/13,670lb (4,280/6,200kg) st SNECMA Atar 9C afterburning turbo-jet
Max speed: 1,460mph (2,350kmh) at 39,375ft (12,000m)
Operational ceiling: 55,775ft (17,000m)
Range: Typical combat radius: 180 miles (290km); III-R: 745 miles (1,200km); III-S, with 2,000lbs (907kg) bomb-load: 808 miles (1,300km)
Armament: 2×30mm DEFA cannon; 2×HM-55 Falcon AAMs (454kg)
Max bomb load: 1×ASM plus 2×1,000lb (454kg) bombs externally

The MD 550 Mirage I flew on 25 June 1955; the larger, much-modified Mirage III prototype which followed it on 17 November 1956 yielded a supremely versatile family of warplanes capable of high- or low-level interception in all weathers, nuclear strike, tactical reconnaissance and strike, combat training, and experimental VTOL and swing-wing derivatives. The first of ten pre-production Mirage III-As, powered by an Atar 9B, and with provision for an SEPR 841 rocket motor, flew on 12 May 1958. The first series-built single-seat model was the interceptor/attack III-C with a similar engine and Cyrano *Ibis* nose radar (first flight 9 October 1960). A 2-seat operational trainer equivalent is the III-B (first flight 21 October 1959). The major production model is the III-E, a longer-range tactical strike version with a more powerful Atar 9C turbojet, optional SEPR 844 rocket motor and Cyrano II. The first production III-E flew on 14 January 1964; trainer equivalent is the III-BE. A tactical reconnaissance version is the III-R, externally recognisable by its snub nose, carrying three oblique and two vertical cameras. The prototype flew on 31 October 1961, and the first production III-R/-RD flew on 1 February 1963. The III-D is a 2-seat version. About 155 III-Cs, 180 III-Es, 50 III-Rs, 20 III-RDs and 40 III-B/-BEs were built for the *Armée de l'Air*. Large numbers of IIIs have been built for, or in, nearly a dozen countries, identified by suffix letters. Customers include Argentina, Australia, Brazil, Israel, Lebanon, Libya, Pakistan, South Africa, Spain, Switzerland, and Venezuela. For ground attack the III-E normally carries two 30mm DEFA cannon in the fuselage, an AS30 air-to-surface missile or two 450kg bombs under the fuselage, and similar-sized bombs, rockets or drop-tanks beneath the wings. For interception, guns are optional, one Matra R530 missile and two Sidewinders being carried. The prototype Mirage 5 flew on 19 May 1967. Based on the III-E airframe and powerplant, it has simplified avionics but seven external stores points giving a greater capacity for fuel and weapon carriage. It can be operated in interceptor or ground attack configuration. Operators include Abu Dhabi, Belgium, Colombia, Libya, Pakistan, Peru, Saudi Arabia, Venezuela and Zaïre.

Hawker Siddeley Buccaneer

Hawker Siddeley Buccaneer S Mk 2A of No 15 Sqdn, RAF, 1972

Span: 44ft 0in (13.41m); folded: 19ft 11in (6.07m)
Length: 63ft 5in (19.33m)
Weight: 62,000lb (28,123kg)
Engine: 2×11,100lb (5,035kg) st Rolls-Royce RB 168-1A Spey Mk 101 turbofans
Max speed: 645mph (1,038kmh) at 200ft (61m)
Operational ceiling: over 40,000ft (9,145m)
Range, typical: 2,300 miles (3,700km)
Armament: None
Max bomb load: 16,000lb (7,257kg) internally and externally

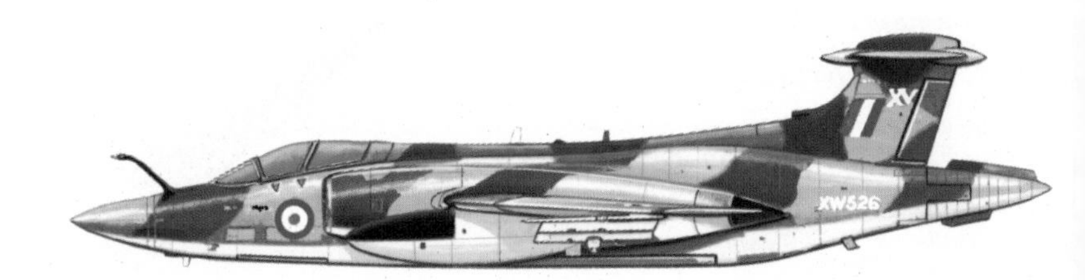

No prototype was ordered of the 2-seat, low-level strike Buccaneer: the initial contract, placed in July 1955, was for 20 B103 development aircraft to meet RN requirement NA39. The first (XK486) flew on 30 April 1958. After carrier trials, the first S Mk 1s were delivered to the RN in 1961, and the first squadron was commissioned in July 1962. Production of the 7,100lb (3,220kg) st Gyron Junior-engined S Mk 1 totalled 40, these equipping Nos 800, 801 and 809 Sqdns and No 736 Training Sqdn FAA. The Buccaneer S Mk 2, first flown on 17 May 1963, had Spey turbofan engines, giving more power, lower fuel consumption, greater accelerating and climbing performance, and a marked increase in range; 84 S Mk 2s were built for the FAA, equipping Nos 800, 801, 803 and 809 Sqdns, entering service in October 1965. The Buccaneer excels in the ground-hugging, under-the-radar penetration role at high subsonic speeds. Its ordnance load includes nuclear weapons or four 1,000lb bombs on its rotatable internal bay. Loads on four underwing pylons include 500, 540 or 1,000lb bombs, Martel anti-radar missiles or 2in, 3in or 68mm rocket pods. A camera pack replaces internal weapons for reconnaissance missions. The SAAF's Maritime Command received 16 S Mk 50s during 1965, basically similar to the S Mk 2; fully 'navalised' although shore based, they have a Bristol Siddeley BS605 twin-barrelled rocket motor in the rear fuselage, giving a 30-second boost of 8,000lb (3,628kg) st to aid take-off. In 1969 the hand-over began of most FAA Mk 2s to the RAF; the first RAF unit to become operational with Buccaneers was No 12 Sqdn at Honington in July 1970 followed by No 208. These aircraft, designated S Mk 2A, have systems and equipment changes to meet RAF requirements. About 70 were converted to 2As and later brought up to 2B standard, with Martels, and a bulged weapons bay door for an additional fuel tank. In addition, 43 new S Mk 2Bs were ordered for the RAF, the first flying on 8 January 1970. First RAF units completely to re-equip with the S Mk 2B were Nos 15 and 16 Sqdns, at Laarbruch in Germany. Three other S Mk 2Bs (XW986-988) were built for special weapons trials by the RAE, and two RAF aircraft were used as trials aircraft for the Panavia Tornado. RN Buccaneers remained in service with No 809 Sqdn in HMS *Ark Royal* until 1978, designated S Mk 2C without Martel-carrying capability, and 2D with Martels.

Grumman A-6 Intruder

Grumman A-6A Intruder of VA-85, US Navy, USS *America*, south-west Pacific area, *circa* 1968

Span: 53ft 0in (16.15m)
Length: 54ft 7in (16.64m)
Weight: 60,626lb (27,500kg)
Engines: 2×9,300kb (4,218kg) st Pratt & Whitney J52-P-8A turbojets
Max speed: 685mph (1,102kmh) at sea level
Operational ceiling: 41,660ft (12,700m)
Typical combat range: 1,920 miles (3,090km)
Armament: None
Max bomb load: 15,000lb (6,804kg) externally

Chosen in late 1957 from 11 designs, the first of eight 2-seat A-6A (originally A2F-1) development aircraft flew on 19 April 1960 with downward-tilting engine tailpipes to give 30 degree thrust deflection to aid take-off; subsequent aircraft have fixed pipes with 7 degree deflection. Replacing Skyraiders on USN attack carriers, it entered service in February 1963, and by mid-1965 was flying in the Vietnam conflict, being well suited to operations in this territory. The Intruder looks deceptively slight for the loads it carries, but is slightly larger than the Buccaneer, fulfilling a similar role with the RN and RAF. For carrier stowage, the A-6's upward-folding wings reduce span to 25ft 4in (7.72m). The four stores stations carry loads including 30 500lb bombs, four pods each containing four 5in Zuni rockets, Bullpup air-to-surface missiles, auxiliary fuel tanks, or nuclear weapons. There is a semi-recessed weapons bay in the fuselage. The initial USN version, the A-6A, was fitted with DIANE (Digital Integrated Attack Navigation Equipment). When A-6A production ended in December 1969, 482 A-6As had been built. Subsequent versions included the EA-6A, A-6B, EA-6B, A-6C, KA-6D and A-6E. The USMC's EA-6A (21 built, plus six converted from A-6As) is an attack escort using ECM to protect strike aircraft, and has a limited strike ability. The A-6B was an A-6A conversion with improved avionics, permitting carriage of anti-radar missiles; the first of 19 flew in 1963. The prototype EA-6B Prowler flew on 25 May 1968. An updated EA-6A with a 3ft 4in (1.02m) longer nose and a 4-man crew, 102 will be ordered. EA models have a large radome on top of the fin. The A-6C, an A-6A conversion, had a ventral turret containing FLIR (forward-looking infra-red) sensors to detect 'difficult' targets, especially at night; 12 A-6As were modified to A-6C standard. The KA-6D is an A-6A variant, 54 of which were converted from A-6As; it first flew on 23 May 1966 and is in use as a carrier-borne 'buddy' refuelling tanker. The A-6E (over 300 ordered) which flew on 27 February 1970 and entered service in 1972, is a further avionics update of the A-6A, with improved weapon delivery. First flown on 22 March 1974, the A-6E TRAM (target recognition attack multi-sensor), carries a ventral fuselage package containing infra-red and laser equipment in addition to standard A-6E avionics.

Northrop F-5

Northrop F-5A of the 10th Fighter Wing (FW), Republic of Korea Air Force, 1965

Span: 25ft 3in (7.70m)
Length: 47ft 2in (14.38m)
Weight: 20,576lb (9,333kg)
Engines: 2×2,720/4,080lb (1,234/1,850kg) st General Electric J85-GE-13 afterburning turbojets
Max speed at all-up weight of 11,450lb (5,193kg): 925mph (1,489kmh) above 36,000ft (11,000m)
Operational ceiling: over 50,000ft (15,250m)
Typical combat radius: 550 miles (885km)
Armament: 2×20mm Colt Browning M39 cannon; provision for 2×Sidewinder AAMs
Max bomb load: 6,200lb (2,812kg)

Four years of indecision and apparent indifference followed the maiden flight, on 30 July 1959, of Northrop's N-156F Freedom Fighter, a compact, nimble and versatile design aimed at the world's smaller air forces which needed modern but inexpensive equipment. Northrop built, as a private venture, three prototypes of the N-156C, in parallel with the N-156T 2-seat trainer version which became the T-38A Talon. The third prototype, an improved model designated N-156F, flew in May 1963, by which time the US Defense Department had selected the single-seat F-5A and 2-seat F-5B for delivery under MAP to several NATO and SEATO air forces. The first production F-5A flew in October 1963. The first foreign deliveries were made to Iran in 1965. Iran eventually received 104 F-5As, 13 RF-5As and 22 F-5Bs, of which some were later re-allocated to South Vietnam, and Pakistan. Recipients include Brazil, Nationalist China, Ethiopia, Greece, Libya, Morocco, Norway, Saudi Arabia, South Korea, the Philippines, Thailand and Turkey. CASA of Spain assembled 36 SF-5As and 34 SF-5Bs for the *Ejército del Aire;* Canadair built 89 CF-5As and 25 CF-5Ds for the CAF, and 75 NF-5As and 30 NF-5Bs for the Netherlands. The F-5A and F-5B have broadly similar performances, the F-5B dispensing with the F-5A's two 20mm Colt-Browning M-39s. Both have provision for up to 6,200lb (2,812kg) of stores on five stations, and wingtip Sidewinder AAMs. In response to the Defense Department's IFA (International Fighter Aircraft) programme for an F-5 successor for MAP countries, Northrop re-engined an F-5B with two 5,000lb (2,268kg) st YJ85-GE-21 turbojets and modified the airframe for improved performance, fuel load and operational versatility. This, designated YF-5B-21, flew on 28 March 1969, and in November 1970, redesignated F-5E Tiger II, was selected. Major differences include a wider fuselage, increased-area wings, with a wing/body airflow strake at the roots, and full-span leading-edge flaps; and an improved fire control system. Airfield performance and manoeuvreability are improved. The F-5E is intended primarily for the air superiority role, with two nose-mounted M-39 20mm cannon and wingtip Sidewinders. It can carry up to 7,000lb (3,175kg) of ordnance, on one under-fuselage and four under-wing stations, in the ground attack role. A production F-5E flew on 11 August 1972, and deliveries began (to South Vietnam) In March 1974. F-5E (and 2-seat F-5F) customers include Brazil, Chile, Iran, Jordan, Kenya, South Korea, Malaysia, Saudi Arabia, Singapore, Sudan, Switzerland, Taiwan, Thailand, the USAF and USN, and Yemen Arab Republic. The RF-5E is a specialised PR version; the proposed F-5G, to replace the F-5E, has 50 per cent more thrust and much-improved performance.

General Dynamics F-111

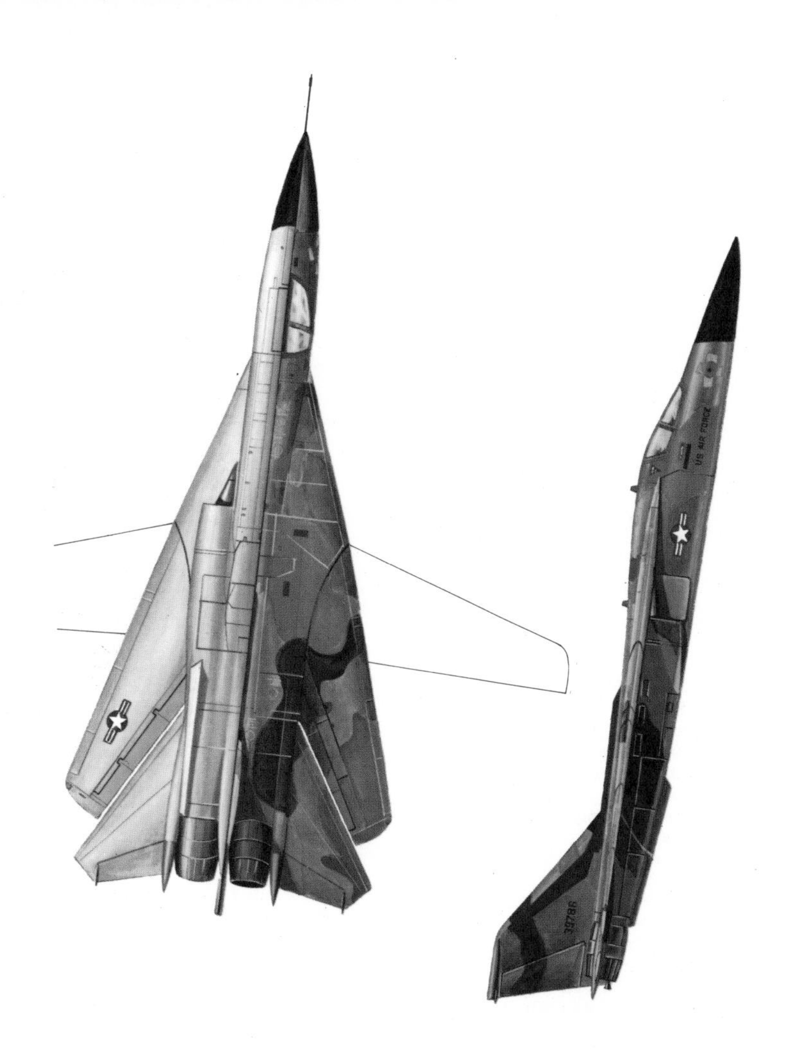

General Dynamics F-111A of Tactical Air Command, USAF, *circa* 1967

Span: 63ft 0in (19.20m); swept: 31ft 11½in (9.74m)
Length: 73ft 6in (22.40m)
Weight: 91,500lb (41,504kg)
Engines: 2×12,500/21,000lb (5,670/9,525kg) st Pratt & Whitney afterburning turbofans
Max speed: 1,650mph (2,655kmh) above 36,000ft (11,000m)
Operational ceiling: over 60,000ft (18,300m)
Range on internal fuel: over 3,800 miles (6,100km)
Armament: 1×20mm M61 Vulcan multi-barrel cannon in internal weapon bay
Max bomb load: 1,500lb (680kg) internally instead of M61 gun; 28,000lb (12,700kg) externally

Winner of the US Defence Department's 1961 Tactical Fighter Experimental competition, the 'swing-wing' F-111 had an extremely unhappy development and early service career. A Senate investigating sub-committee, reporting in 1971, found that five major errors of judgement had been made. The cost of acquiring about 500 F-IIIs exceeded that estimated for the 1,700 originally required – and that less than 100 approached the required standard. Originally, 18 F-111As for the USAF and five F-111Bs, with a shorter nose and extended-span wings, for the USN, were procured for evaluation. However, the naval version was overweight for its carrier-based fleet defence fighter role, and that programme was cancelled in mid-1968 after two additional examples had been completed. Evaluation of the F-111A, flown on 21 December 1964, was followed by 141 production examples, delivery beginning in 1967. Within five days of their first combat sorties in Vietnam, two were lost. A reconnaissance version designated RF-111A flew on 17 December 1967. Two F-111Ks from a cancelled British requirement for 50 were assigned to the USAF as YF-111As for experimental duties. The F-111A, which became subject to high-speed, high-altitude flight restrictions, was superseded by the F-111E with modified air intakes (94 built) and the F-111D with improved avionics and TF30-P-9s (96 built). SAC's requirement for 210 FB-111A 2-seat bombers, was reduced to 76, delivery beginning in October 1969. The two prototypes were converted F-111As, the first being flown as an FB-111A on 30 July 1967. The FB-111A had Mk IIB avionics, the F-111B's longer-span wings, TF30-P-7s and a maximum warload of 50 750lb bombs (two internally, the rest on eight underwing pylons). Delivery of 24 similar F-111Cs, with TF30-P-3s and Mk I avionics, began in 1973 to the RAAF for strike duties. Following the F-111E for the USAF were 82 F-111F fighter-bombers, combining the F-111E's and A's best features with more powerful TF30-P-100s, giving a considerably enhanced performance and capability. As this version has shown, the General Dynamics F-111 is a fully-viable combat aircraft, and its unfortunate career has been more due to mismanagement than to fundamental design defects. It was the first combat aircraft produced with variable-geometry wings, and the first tactical fighter designed to satisfy the joint needs of the USAF and the USN. The wings have 16 degrees of sweep fully-foward, and 72.5 fully-swept. External loads can be carried beneath the fuselage, the fixed portion of the wings, and the outer wings on pylons which pivot, as the wings sweep, to align fore and aft. The important new role of ECM tactical jamming is assigned to the EF-111A, distinguished by its large fin-tip antenna pod and bulged weapons bay. It first flew on 10 March 1977, and the first of 42 (to be converted by Grumman from F-111Fs) was delivered to the USAF in 1981.

British Aerospace (Hawker Siddeley) Harrier

Hawker Siddeley Harrier GR Mk 1 of the RAF, flown in the 1969 *Daily Mail* trans-Atlantic air-race by Sqdn Ldr T Lecky-Thompson, setting a world record of 5hr 57min for the 3,490 mile (5,615km) flight between the centres of London and New York

Span: 25ft 3in (7.70in)
Length: 45ft 6in (13.87m)
Weight: over 25,000lb (11,340kg)
Engine: 19,000lb (8,620kg) st Rolls-Royce Pegasus Mk 101 vectored-thrust turbofan
Max speed: over 737mph (1,186kmh) at low altitude
Operational ceiling: over 50,000ft (15,240m)
Range with one in-flight refuelling: over 3,455 miles (5,560km)
Armament: 2×30mm Aden cannon in underfuselage pods
Max bomb load: 5,000lb (2,268kg) of bombs, rockets, gun pods etc externally

The Harrier is the world's first fixed-wing operational combat aircraft with V/STOL capabilities. Originating as a private venture, the first of six P1127 prototypes made its initial vertical take-off on 21 October 1960, and its first transition from hovering to forward flight some 11 months later. Hawker Siddeley built nine Kestrel F(GA) Mk 1s (first flown 7 March 1964), which were thoroughly tested in Britain by RAF, *Luftwaffe*, and US Army, Navy and Air Force pilots in 1965. In February 1965 production was authorised of six Harrier development aircraft for the RAF (first flight 31 August 1966). The GR Mk 1 entered service with Air Support Command on 1 April 1969, followed by the T Mk 2 in 1970. In May 1969 four Harriers flew in the *Daily Mail* trans-Atlantic air race. The 2-seat Harrier differs from the single-seater in having a longer nose, accommodating two cockpits in tandem, enlarged fin and lengthened tail cone. Both models have the same weapons capability, with attachment points under the wings (four) and fuselage (two), for gun pods, bombs, rockets and flares; a reconnaissance-camera pod can be carried on the fuselage centreline pylon. There is no built-in armament. RAF Harriers are cleared for operation with a maximum load of 5,000lb. They have undergone progressive in-service improvement, first to GR Mk 1A/T Mk 2A standard by installing the 20,000lb (9,072kg) st Pegasus 102 in place of the 19,000lb (8,618kg) 101, and to GR Mk 3/T Mk 4 standard by replacing the 102 by the 21,500lb (9,752kg) st Pegasus 103. A laser rangefinder for the GR Mk 3 is installed in a 'thimble' nose fairing. The Pegasus 103 (US designation F402-RR-401) is installed in the Harrier Mk 50 export version, ordered by the USMC as the single-seat AV-8A (102) and 2-seat TAV-8A (eight). The first ten AV-8As were delivered in 1971, initially with Pegasus 102s, but were later re-engined with Pegasus 103s. The Spanish Navy acquired, through the USN, 11 AV-8Ss and two TAV-8Ss. The FAH's Sea Harrier FRS Mk 1 (24 ordered) has a redesigned nose with a raised cockpit, and a 9,525kg (21,000lb) st Pegasus 104. The first squadron became operational in 1980. Sea Harriers will equip the RN's *Invincible*-class light carriers, which are fitted with the 'ski-jump' launch system. The Indian Navy has ordered six similar FRS Mk 51s. An improved AV-8B version, with double the payload/range capability, is being developed by McDonnell Douglas; meanwhile the USMC AV-8As are being upgraded to interim improved standard as AV-8Cs.

MiG-25 (Foxbat)

Mikoyan MiG-25R (Foxbat-B) of the Soviet Air Force, *circa* 1972

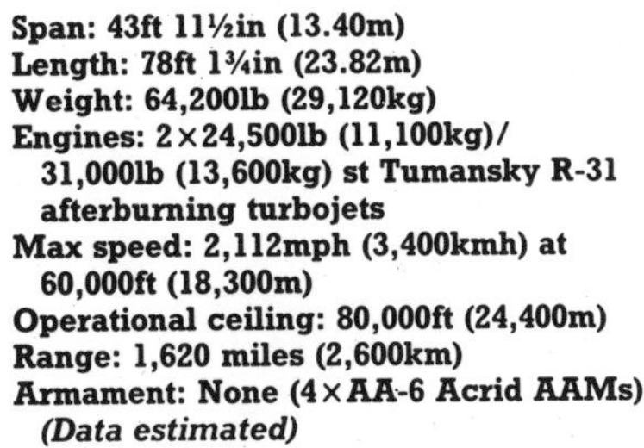

Span: 43ft 11½in (13.40m)
Length: 78ft 1¾in (23.82m)
Weight: 64,200lb (29,120kg)
Engines: 2×24,500lb (11,100kg)/ 31,000lb (13,600kg) st Tumansky R-31 afterburning turbojets
Max speed: 2,112mph (3,400kmh) at 60,000ft (18,300m)
Operational ceiling: 80,000ft (24,400m)
Range: 1,620 miles (2,600km)
Armament: None (4×AA-6 Acrid AAMs)
(Data estimated)

In 1965 the USSR received official confirmation of a new world speed record over a 1,000km (621 mile) closed circuit by an aircraft which it identified as an E-266. In 1967 at the Aviation Day display at Domodedovo, western observers first saw this large twin-engined, single-seat military aeroplane, which the commentator described as 'high altitude all-weather interceptor' with a Mach 3 performance. Contemporary with the MiG-23, the MiG-25 was designed for reconnaissance and high-altitude interception as a counter to America's SR-71A strategic reconnaissance aircraft and XB-70 Valkyrie strategic bomber, both capable of Mach 3 flight. Further evidence of the Foxbat's capabilities was given on 5 October 1967, when a 2,000kg payload was carried to a height of 98,349ft (29,997m) and a 500km (310 mile) closed-circuit speed of 1,852.61mph (2,981.5kmh) was attained on further record flights; and on 27 October 1967, when the 1,000km record of 1965 was raised to 1,814.81mph (2,920.67kmh). It took a further crop of records in 1972. Such achievements confirmed the Foxbat's very high performance; yet not until 1971 did firm evidence begin to appear regarding its service status, and, even some ten years after its design was initiated, much information regarding the MiG-25s combat potential remained speculative. It was described by one US Secretary of the Air Force as 'probably the best interceptor in production in the world today'. The original MiG-25 interceptor had only missile armament; the later 2-seat MiG-25MP has internal cannons and 'look-down' radar, and increased endurance. A 2-seat trainer has also been produced. The interceptor has four underwing hardpoints for air-to-air missiles, probably of the 'snap-down' type. The MiG-25R for high altitude reconnaissance is distinguished by a slightly longer, camera-packed nose (with no di-electric covering over the nose radar) and tail-fins which have broken taper on their trailing edges. It is unarmed, although the underwing hardpoints are retained. The first reports of operational MiG-25s appeared in spring 1971, when Soviet Air Force squadrons equipped with them were airlifted into Egypt. These were later withdrawn, but Foxbats were later reported in other Middle Eastern theatres of conflict. In September 1976, a defecting Soviet pilot landed his MiG-25 in Japan, giving Western experts the opportunity to examine it before it was returned. Designed as a single-seater, construction is mainly of nickel steel alloy, with some titanium.

SAAB 37 Viggen

SAAB AJ 37 Viggen of F7 Wing, Swedish Air Force, Såtenäs, 1972

Span: 34ft 9¼in (10.60m)
Length (including probe) 53ft 5¾in (16.30m)
Weight (normal armament): approx 35,275lb (16,000kg)
Engine: 14,770/26,015lb (6,700/11,800kg) st Volvo Flygmotor RM8A afterburning turbofan
Max speed: 1,320mph (2,125kmh) at 39,375ft (12,000m)
Tactical radius (high-low-high) with external armament: 620 miles (1,000km)
Armament: 3 underfuselage and 4 underwing points for RB04E or RB05A, RB75 (Maverick) ASMs, RB24 (Sidewinder) or RB28 (Falcon) AAMs, 30mm Aden gun pods, Bofors 135mm rockets etc

The Viggen (Thunderbolt) replaced SAAB's Lansen and Draken in attack and reconnaissance roles. SAABs canard layout, with its anhedral mainplane and smaller, no-dihedral, flap-equipped foreplane, is a bold but logical design; otherwise the Viggen resembles the Phantom, which the Swedish studied closely before pursuing the native design. Developed during 1952-58 by a team led by Erik Bratt (also responsible for the Draken), the design was not 'frozen' until May 1963. The canard layout, a variation on the Draken's 'double-delta', enables the Viggen to retain the same excellent STOL performance as its predecessor while operating at significantly greater combat weights at all levels with a superior rate of climb. The first of seven prototype/pre-production Viggens flew on 8 February 1967. The first six, all single-seaters, were flying by April 1969, followed on 2 July 1970 by the seventh, the SK37 2-seat operational trainer prototype. The first production Viggen flew on 23 February 1971. Orders totalled 175 AJ37 all-weather attack aircraft with secondary, interceptor capability, and the SK37. The development and service introduction programme proceeded so well that the same funds covered an additional five aircraft. The AJ37 entered service in mid-1971, replacing the A32A Lansen; SK37 deliveries began in June 1972. Later versions include the interceptor/attack JA37; the tactical reconnaissance SF37 (first flown on 21 May 1973), to replace the S32C Lansen; and the armed sea surveillance SH37, to replace the S35E Draken. To power the Viggen, Volvo Flygmotor AB developed the Pratt & Whitney JT8D-22 civil turbofan to deliver 26,450lb (12,000kg) thrust. RB04 or RB05 air-to-surface missiles are the primary armament in the attack role. The seven permanent stores points beneath the fuselage and wings (and two optional underwing) can carry air-to-surface rockets, bombs, 30mm Aden guns or mines; AAMs for interception; or camera or radar pods for reconnaissance. A miniaturised digital computer provides automated navigation, target location and fire control, integrated with Sweden's STRIL 60 ground defence network; it can also fly the aeroplane automatically. Like its predecessors, it operates from Sweden's under-ground hangar system and can take-off from any 500m (1,640ft) stretch of straight trunk road.

MiG-23 (Flogger)

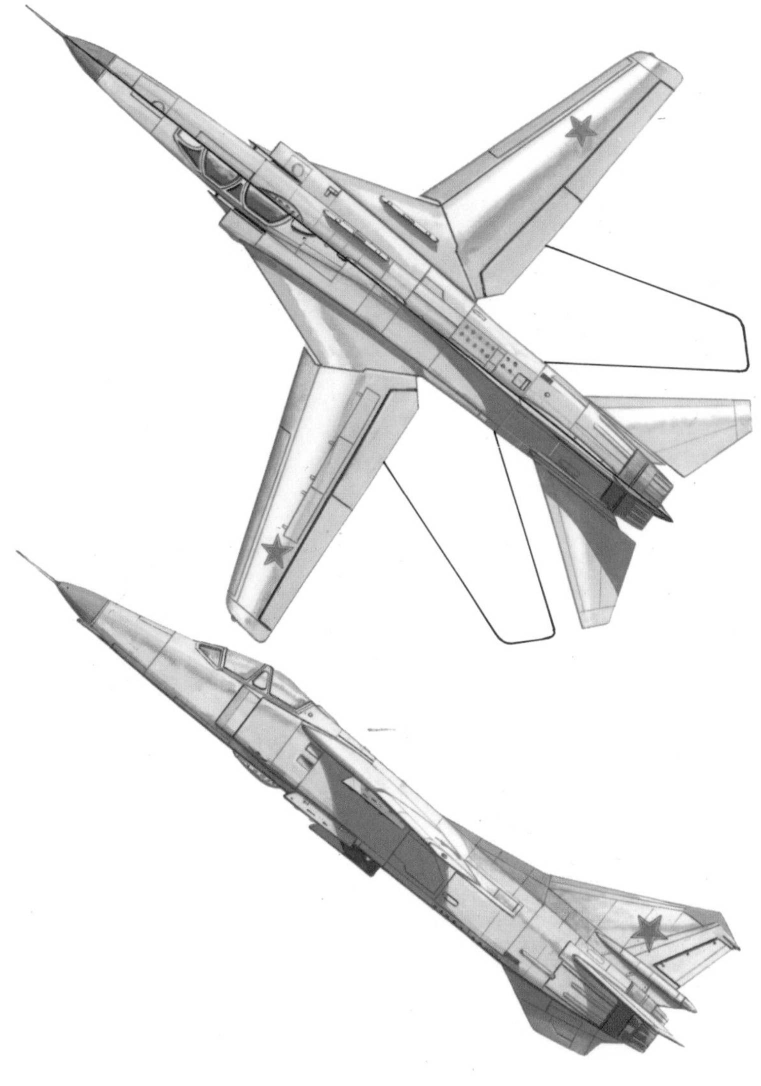

Mikoyan MiG-23U (Flogger-C) of the Soviet Air Force, 1972-73

Span: 46ft 9in (14.25m); swept: 26ft 9½in (8.17m)
Length: 55ft 1½in (16.80m)
Weight: 33,070lb (15,000kg)
Engine: 17,637/25,353lb (8,000/11,500kg) st Tumansky R-29B afterburning turbofan
Max speed: 1,520mph (2,445kmh) above 36,000ft (11,000m)
Operational ceiling: 61,025ft (18,600m)
Combat radius: 600 miles (960km)
Armament: 1×twin-barrel 23mm GSh-23L cannon
Max bomb load: 5 external points for AA-7 Apex or AA-8 Aphid AAMs, rockets or other stores
(Data estimated)

At the 1967 Aviation Day display at Domodedovo Airport, near Moscow, examples were shown of two single-seat prototype aircraft featuring variable-geometry or 'swing' wings: the Sukhoi Su-7E and Mikoyan MiG-23. Both entered service, the former as the 2-seat Su-17 (Fitter-B) and the latter as the single-seat MiG-23 MF (Flogger-B) and the 2-seat MiG-23U (Flogger-C). Powered by a single after-burning turbojet, the Flogger-A prototype was credited with a Mach 2.2 capability at medium and high altitudes, and bore a general resemblance in size and configuration to the contemporary French Dassault Mirage G-01 prototype. The MiG-23 entered production in 1970-71, and subsequently, it was reported in Egypt, East Germany and Syria. In 1974 photographs of the MiG-23U enabled a more up-to-date appraisal to be made. Slightly smaller than was at first thought, and probably with a less powerful engine, its wing has 21 degrees of leading-edge sweep in the fully-forward position, increasing to 72 degrees when fully swept, and, unusual on variable-geometry aircraft, has full-span trailing edge flaps. The tailplane is all-moving. Ample keel area is provided by a large main fin and a dorsal fairing (larger still on the MiG-23U), and by a ventral fin beneath the tailpipe, the lower portion of which folds starboard to give ground clearance. There are two lateral airbrakes below the tailplane, and three attachments for external stores under the fuselage and one beneath the wing glove box on each side. Ahead of the fuselage hardpoints a fairing houses a fixed, twin-barrel 23mm GSh-23 gun for close-range fighting. The MiG-23 series all-weather and tactical interceptor represents the fourth generation of Soviet jet fighters and unlike previous generations has sophisticated weaponry, ECM and navigation systems. There are six operational versions: the first production all-weather interceptor with radar nose, the MiG-23MF, carrying a GSh-23 and missiles; the simplified export Flogger-E version; the MiG-23U Flogger-E 2-seat trainer; and a modified B with small dorsal fin and under-nose sensors, known as Flogger-G. The MiG-27 (Flogger-D) is a specialised ground-attack tactical fighter with a 23mm five-barrel rotary gun and missiles and laser designator replacing the interceptor's radar. An export version, combining features of both the MiG-23 and -27, is designated MiG-23BM (Flogger-F). The MiG-23MF serves with PVO *Strany* (the National Air Defence Force) and *Frontovaya Aviatsiya* (Frontal Aviation), which supports the Soviet Army and also operates the MiG-27. The MiG-23 has been exported, like the MiG-15 and MiG-21, to both communist and non-communist countries.

SEPECAT Jaguar

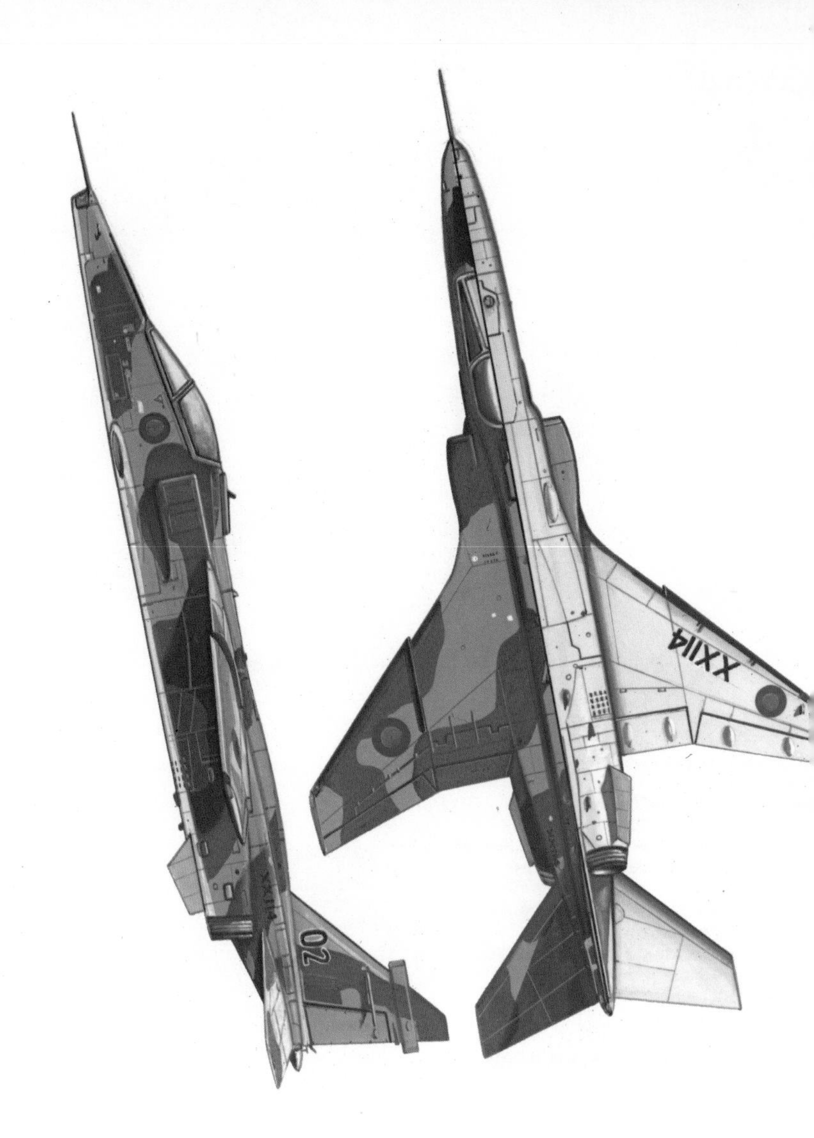

SEPECAT Jaguar GR Mk 1 of the RAF Jaguar Operational Conversion Unit, Lossiemouth, 1973

Span: 28ft 6in (8.69m)
Length (excluding nose probe): 50ft 11in (15.52m)
Weight: 34,000lb (15,422kg)
Engines: 2×5,115/7,304lb (2,320/3,313kg) st Rolls-Royce Turboméca Adour Mk 102 afterburning turbofans
Max speed: 990mph (1,593kmh) above 36,000ft (11,000m)
Combat radius (high-low-high mission) on internal fuel: 507 miles (815km)
Armament: 2×30mm Aden cannon
Max bomb load: 10,000lb (4,536kg) externally

By no means the 'poor man's Phantom' that some claim, the Franco-British Jaguar is the first fixed-wing combat aircraft to be designed, developed and manufactured on a fully-international basis. The airframe is the product of collaboration between the French Dassault-Breguet group and British Aerospace, the powerplant is the Rolls-Royce/Turboméca Adour turbofan. Overall responsibility for the programme is vested in the Société Européenne de Production de l'Avion ECAT (SEPECAT), formed in May 1966, a year after the signing of a project agreement between the French and British Defence Ministries. Development was begun of three basic versions: a single-seat tactical support version (Jaguar A for the *Armée de l'Air,* first flight 29 March 1969, and S for the RAF, first flight 12 October 1969); a single-seat carrier-based tactical version (M for the *Aéronavale,* first flight 14 November 1969); and 2-seat advanced (E for France) or operational (B for Britain) trainers. Eight flying prototypes were built: two each of the A, E and S and one each of the M and B. First to fly, on 8 September 1968, was the E-01 prototype; the eighth (the B-08) first flew on 30 August 1971. Development of a naval version ceased in 1973, but renewed interest was indicated in 1980. The Anglo-French agreement called for 200 for each country. RAF designations are GR Mk 1 for the S and T Mk 2 for the B. All versions have a fixed armament of two 30mm DEFA 553 or Aden cannon in the fuselage underside, except the B, which has one Aden. The tactical versions have one under-fuselage and four underwing weapon stations, carrying a maximum of 10,000lb (4,500kg) of bombs, rockets, Martel anti-radar missiles or drop-tanks. Jaguar A can carry an AN52 tactical nuclear weapon. Wingtip AAMs (Sidewinders or later types) can be fitted, and the single-seat versions have in-flight refuelling. The 2-seat versions have a secondary strike capability with similar weapon loads. Production deliveries began in 1973, France's *7e Escadre de Chasse* being first recipient (A and E), and the RAF received GR Mk 1s for ground crew and flying training at the OCU at Lossiemouth. First operational RAF Jaguar squadron, No 54, was formed on 29 March 1974, followed by No 6 Sqdn and six other operational Jaguar squadrons. A Jaguar International, with uprated engines, increased weapon load and other options, has been exported to Ecuador (12), Oman (24) and India (40, with options for 45 more).

Grumman F-14 Tomcat

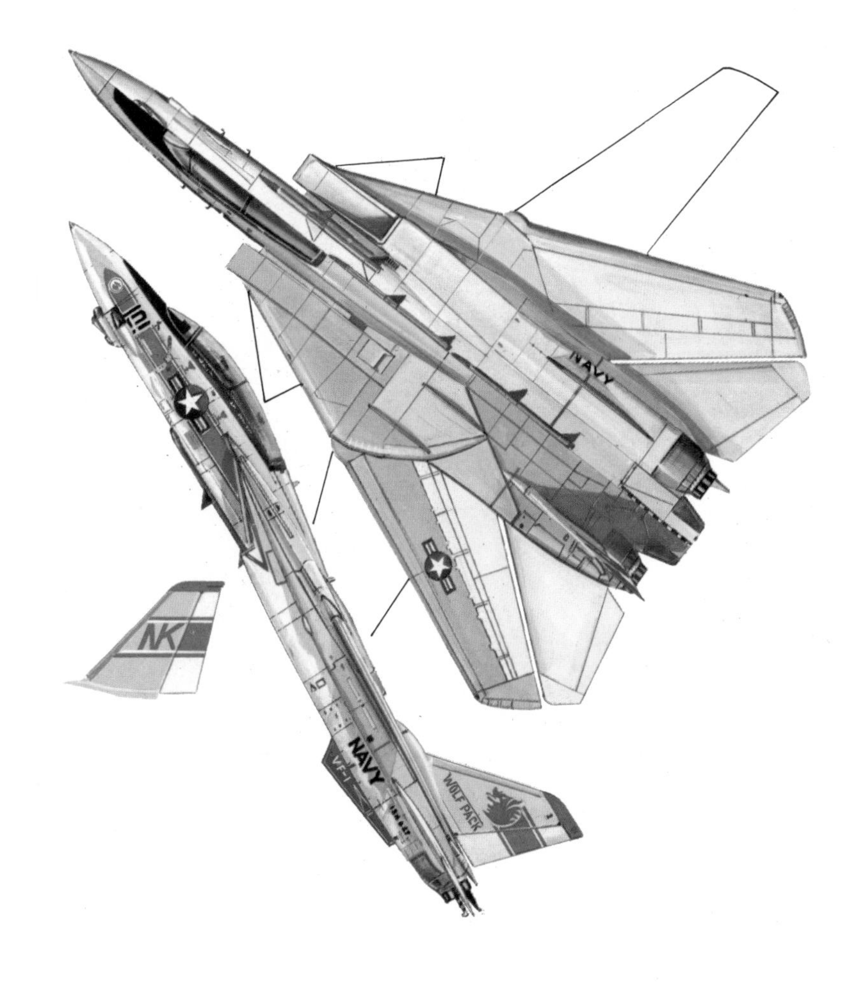

Grumman F-14A Tomcat of VF-1, US Navy, US Naval Air Station Miramar, California, 1973

Span: 64ft 1½in (19.54m); wings swept: 38ft 2½in (11.65m)
Length: 61ft 11 9/10in (18.89m)
Weight: 74,348lb (33,724kg)
Engines: 2×20,900lb (9,480kg) st Pratt & Whitney TF30-P-412A afterburning turbofans
Max speed: over 1,450mph (2,333kmh) above 36,000ft (11,000m)
Operational ceiling: above 50,000ft (15,240m)
Typical combat radius: 450 miles (724km)
Armament: 1×20mm M61A1 Vulcan six-barrel cannon; 4×Phoenix or Sparrow AAMs under fuselage, and/or 4× Sidewinders or 2×Phoenix or Sparrows under wings
Max bomb load: 14,500lb (6,577kg) of bombs and/or missiles, ECM pods etc

The Tomcat, developed in place of the cancelled F-111B for the USN, was declared winner over four other competitors in a Navy design programme in January 1969 and is a 2-seat, multi-purpose carrier-based fighter. The first of 12 development aircraft flew on 21 December 1970, but crashed on the landing approach after its second test flight nine days later. A second F-14 flew on 24 May 1971. The USN has planned procurement of 521 F-14As; about 400 were in service by 1981. Pre-revolutionary Iran received 80. Avionics and other features are based on those already developed by Grumman for the F-111B, and landing gear on that of the A-6 Intruder. The F-14A has a fixed armament of one M61-A1 Vulcan 20mm multi-barrel cannon, with 675 rounds, in the port side of the lower front fuselage. There are recessed stations under the fuselage, and pylons beneath the fixed portion of the wings, on which can be carried six Phoenix and two Sidewinder, or four Sparrow and four Sidewinder, air-to-air missiles; or drop tanks. The F-14A's computer-controlled radar and weapon systems can engage up to six aircraft at one time, and can detect targets at 185km (115 miles), although, ironically, most sightings are visual. The Tomcat can also be operated in the attack role, carrying bombs or bomb-and-missile combinations up to a maximum external load of 14,500lb (6,577kg). The F-14A is unique among variable-geometry ('swing-wing') aircraft in having, in addition to variable-sweep wings, small movable foreplane surfaces housed inside the leading-edge roots of the fixed portions of the wings. Fulfilling a similar function to the 'moustaches' of the Dassault Milan, these can be extended forward into the airstream as the main wings swing backward, controlling changed in the centre-of-pressure position. The wings themselves have 20 degrees of sweep when fully forward and 68 degrees when fully back. The first operational USN Tomcat unit was VF-1; with VF-2, this squadron flew the first operational F-14A sorties, from the USS *Enterprise,* in March 1974. A carrier wing normally comprises two F-14 squadrons. By 1980 16 USN squadrons operated the F-14; when not at sea they are shore-based at NAS Miramar, California, and NAS Norfolk, Virginia. Engine problems in service required modifications to the TF30 turbofans.

McDonnell Douglas F-15 Eagle

McDonnell F-15A Eagle, development aircraft, 1973

Span: 42ft 9¾in (13.05m)
Length: 63ft 9in (19.43m)
Weight: 56,000lb (25,401kg)
Engines: 2×25,000lb (11,340kg) st (approx) Pratt & Whitney F100-PW-100 afterburning turbojets
Max speed: over 1,650mph (2,655kmh) above 36,000ft (11,000m)
Operational ceiling: 66,900ft (20,390m)
Range with FAST packs: more than 3,450 miles (5,560km)
Armament: 4×Sidewinder and 4×Sparrow AAMs and 1×20mm M61A1 six-barrel cannon
Max bomb load: 16,000lb (7,257kg) externally

McDonnell Douglas's F-15 design was declared the winner of a USAF competition over proposals from Fairchild Hiller and North American Rockwell, in December 1969 for a new air superiority fighter to counter the ascendancy of Soviet fighters in this category, particularly the MiG-25. The Eagle is a single-seat, twin-turbofan all-weather fighter, armed with a built-in rapid-firing gun (initially the 20mm M61-A1, with 1,000 rounds) and four Sparrow and four Sidewinder AAMs. It has a secondary attack capability and a radar capable of detecting low-flying targets, but the basic roles are air-to-air interception, fighter sweep, escort and combat air patrol. Particular emphasis has been placed on the dog-fighting aspects of manoeuvrability and acceleration – an area in which the gap between US and the latest Soviet fighters had widened appreciably. Initial funding covered the production of 20 development aircraft (including two 2-seaters, and the first of these (71-0280) flew on 27 July 1972, followed by the first 2-seat TF-15 on 7 July 1973, which has a similar combat capability, encompassing reconnaissance and attack roles, and was thus redesignated F-15B. The first Eagles were delivered to USAF TAC in November 1974. The USAF plans an eventual purchase of 749; about two thirds of these had been delivered by 1981. Those delivered since mid-1979 are F-15Cs (single-seat) and F-15Ds (2 seat) with extra internal fuel, plus provision for FAST (Fuel And Sensor Tactical Packs – pallets for fuel and reconnaissance/ECM gear which fit flush against the intake trunks. Five weapons stations on the standard F-15 permit carriage of up to 16,000lb of ordnance; additional weapons can be attached to the FAST packs. The Strike Eagle 2-seat attack version has advanced nav/attack radar, Pave Tack night/all-weather targeting pods, with infra-red tracking and laser-spot designation, and up to 24,000lb of weapons. USAF Eagles were serving in 1981 with the 1st, 33rd and 49th TFW, and two training wings, in the USA; the 36th TFW and 32nd TFS in Europe, and part of the 18th TFW in Okinawa. Foreign customers include Israel (25), Japan (100) and Saudi Arabia (60). Ironically, the light-weight, 'low-cost', highly maneouvrable, Mark 2 General Dynamics F-16 air superiority fighter was developed to counter the F-15A's sophistication. The F-15 and F-16 provide the basis of TAC through the 1980s, but the F-15 will be most numerous.

Abbreviations

AAEE Aeroplane and Armament Experimental Establishment (UK)
AAM Air-to-air missile
ADC Aerospace Defense Command (USAF)
AEF American Expeditionary Force
AI Airborne interception radar
AMI *Aeronatucia Militare Italiana*
ANG Air National Guard (USA)
ASR Air/sea rescue
ASV Air to surface vessel radar
AVG American Volunteer Group
A-VMF *Aviatsiya-Voenno Morskoi Flot* (Soviet Naval Aviation)
BS; BG; BW Bomber Squadron; – Group; – Wing (USAF/USAAF)
BSh Bronirovannyi Shturmovik (armoured assaulter = Il-2; USSR)
CAF Canadian Armed Forces
COIN Counter-insurgency
CR *Caccia Rosatelli* (Fiat fighters; Italy)
ECM Electronic countermeasures
F Fighter (UK designation); *or: Forsirovanny* (increased power; USSR designation)
FAA Fleet Air Arm (UK)
FAFL *Forces Aériennes Françaises Libres*
FB Fighter bomber (UK designation)
F(GA) Fighter (ground attack) (UK designation)
FMA Fábrica Militar de Aviones (Argentina)
FR Fighter-reconnaissance (UK designation)
FS; FG; FW Fighter Squadron; – Group; – Wing (USAF/USAAF)
FZG *Fernzielgerät* (remote aiming device; Germany)
GR General reconnaissance (UK designation)
GST *Gidro-samolet transportny* (seaplane transport = Soviet PBY)
JAAF Japanese Army Air Force
JASDF Japan Air Self-Defence Force
Jasta *Jagdstaffel* (fighter squadron; Germany)
JG *Jagdgeschwader* (fighter wing; Germany)
JGSDF Japan Ground Self-Defence Force
JMSDF Japan Maritime Self-Defence Force
JNAF Japanese Navy Air Force
KG *Kampfgeschwader* (battle or bomber wing; Germany)
M *Modifikatsirovanny* (= modified; USSR)
MAC Manufacture d'Armes de Châtellerault (armament; France)
MAD Magnetic Anomaly Detection
MAP Military Assistance Program (USA)
NACA National Advisory Committee for Aeronautics (USA)
NAS Naval Air Station (UK, USA)
NEI Netherlands East Indies
NF Night fighter (UK designation)
NJG *Nachtjagdgeschwader* (night fighter wing; Germany)
NR Nudelman-Richter (armament, USSR)
P Pursuit (USA designation); *or: Perekhvatchik* (interceptor; Soviet designation)
PAF Pakistan, or Polish, Air Force
PR Photographic reconnaissance (UK designation)
PVO-*Strany* *Protivovozdushnaya oborona* (Protective air defence – homeland; USSR)
RAAF Royal Australian Air Force
RAE Royal Aircraft Establishment (UK)
RAF Royal Air Force (UK); *or:* Royal Aircraft Factory (UK)
RAN Royal Australian Navy
RCAF Royal Canadian Air Force
RFC Royal Flying Corps (UK)
RLM Reichsluftfahrtministerium (German Aviation Ministry)
RN Royal Navy (UK)
RNAS Royal Naval Air Service (UK)
RNZAF Royal New Zealand Air Force
RP Rocket projectile
RS *Rakety Snaryad* (rocket projectile; USSR)
RSI Republica Sociale Italiana
SAAF South African Air Force
SAC Strategic Air Command (USAF)
SAR Search and rescue
SEAC South-East Asia Command
SEPR Société d'Etude de la Propulsion par Réaction (France)
ShKAS Shpitalny-Komaritsky Aviatsionny Skorostrelny (armament; USSR)
SPS *Sduva Pogranichnovo Sloya* (boundary layer control system; USSR)
Srs Series
StG *Stukageschwader* (dive-bomber wing; Germany)
SWPA South-West Pacific Area
TAC Tactical Air Command (USAF)
TAF Tactical Air Force
TF Torpedo fighter (UK designation)
TsKB Tseutralnoz Konstruktorskoe Byuro (Central Design Bureau)
UAR United Arab Republic
USAAC United States Army Air Corps
USAAF United States Army Air Force
USAF United States Air Force
USAFE United States Air Force in Europe
USCG United States Coast Guard
USMC United States Marine Corps
USN United States Navy
UTI *Uchnebo-trenirovochny istrebitel* (fighter trainer; USSR)
V *Versuchs* (experimental; Germany)
VA Attack Squadron (USN)
VB Bomber Squadron (USN)
VF Fighter Squadron (USN)
VMF Marine Fighter Squadron (USMC)
VMT Marine Training Squadron (USMC)
VP Patrol Squadron (USN)
VS Scout Squadron (USN)
VT Torpedo Squadron (USN)
V-VS *Voenno-vozdushnye sily* (Military Air Forces; USSR)
Vya Volkov-Yartsev (armament; USSR)
WNF Wiener-Neustädter Flugzeugwerke (Austria)
ZG Zerstorergeschwader (destroyer aircraft or 'heavy fighter' wing; Germany)

Index

Names of manufacturers are in **bold** type, types of aircraft or engines in *italics;* page numbers in **bold** type refer to illustrations and in *italic* to specifications.